Lecture Notes in Control and Information Sciences

Edited by A.V. Balakrishnan and M. Thoma

9

Yousri M. El-Fattah
Claude Foulard

Learning Systems: Decision, Simulation, and Control

Springer-Verlag
Berlin Heidelberg New York 1978

Authors
Dr. Y. M. El-Fattah
Electronics Laboratory
Faculty of Sciences
Rabat, Marocco

Professor C. Foulard
Automatic Control Laboratory
Polytechnic Institute of Grenoble
Grenoble, France

ISBN 3-540-09003-7 Springer-Verlag Berlin Heidelberg NewYork
ISBN 0-387-09003-7 Springer-Verlag NewYork Heidelberg Berlin

Printed in Germany

Printing and binding: Beltz Offsetdruck, Hemsbach/Bergstr.
2061/3020-543210

FOREWORD

This monograph studies topics in using learning systems for decision, simulation, and control. Chapter I discusses what is meant by learning systems, and comments on their cybernetic modeling. Chapter II concerning decision is devoted to the problem of pattern recognition. Chapter III concerning simulation is devoted to the study of a certain class of problems of collective behavior. Chapter IV concerning control is devoted to a simple model of finite Markov chains. For each of the last three chapters, numerical examples are worked out entirely using computer simulations. This monograph has developed during a number of years through which the first author has profited from a number of research fellowships in France, Norway, and Belgium. He is grateful to a number of friends and co-workers who influenced his views and collaborated with him. Particular thanks are due to W. Brodey, R. Henriksen, S. Aidarous, M. Ribbens-Pavella, and M. Duflo.

Y.M. El-Fattah
C. Foulard

C O N T E N T S

ABSTRACT

This monograph presents some fundamental and new approaches to the use of learning systems in certain classes of decision, simulation, and control problems.

To design a learning system, one should first formulate analytically the goal of learning that has to be reached at the end of the learning process. As a rule that goal of learning depends on the environment input-output characteristics - conveniently considered to be stochastic - for which there is not sufficient a priori information. That incomplete definition of the goal of learning is compensated by necessary processing of current information. Basic definitions and concepts related to learning systems are presented in Chapter I.

As for decision problems we consider the class of pattern recognition problems in Chapter II. Learning systems can be trained to apply optimum statistical decision algorithms in the absence of a priori information about the classified patterns. The accompanying problem of feature extraction in pattern recognition is also discussed. As an application, we consider the problem of optimal measurement strategies in dynamic system identification. Numerical results are given.

In Chapter III we present a novel model of learning automata for simulating a certain class of problems of collective behavior. Two applications are considered. One is the resource allocation, and the other the price regulation in a free competitive economy. The model performance is studied using computer simulations. Analytical results concerning the limiting behavior of the automata are also given.

A certain control problem of stochastic finite systems modelled as Markov chains is considered in Chapter IV. The control decision model is considered to be a learning automaton which experiments control policies while observing the system's state at the consecutive epochs. Two cases are studied : complete and incomplete a priori information. In the latter case the automaton's policy is dual in nature - in the sense that it identifies the chain's transition probabilities while controlling the system. Conditions are given for the control policy to converge with probability 1 to the optimal policy. Acceleration of the adaptation process is also examined. Computer simulations are given for a simple example.

CHAPTER I

CYBERNETICS OF LEARNING

"Lacking a birth in disorder,
The enlivining detestation of order,
No liberating discipline can ever see
or be the light of a new day."
David Cooper, "The Grammar of Living."

1.1 SYSTEM CONCEPT.

Various definitions can be given for a system. A definition isclearly dependent on the context which it intends to serve. For our purposes the system is defined by behavior, i.e. by the relationship between its input and output.

A model is adequate to describe a system's behavior when it simulates the relationship between the system's output and input. In other words the model for any given sequence of inputs produces the same sequence of outputs as the system. The input x of the system at any time instant is assumed to belong to the set of possible alternatives X. The output y, likewise, belongs to the set of possible alternatives Y. It is usually assumed that X and Y contain a finite number of elements (vectors).

Different types of systems can be distinguished depending on their behavior or kind of relationship between their inputs (stimuli) and outputs (responses). A clasification of certain types is given below.

(a) Deterministic Systems. All the relations are presented by mappings (either one-to-one or many-to-one). In other words, the output variables are functions of the input variables. No probabilities have to be assigned to elements of the relations. Deterministic systems can be subdivided into :

i - Combinational (memoryless) systems. The output components are uniquely determined as certain combinations of the instantaneous values of the input components.

ii - Sequential systems. In this case there exists at least one input which is associated with more than one output. The different outputs of the system to the same input belong to different, but accurately defined, sequences of inputs which preceded the given input.

(b) Probabilistic (Stochastic) Systems. At least one of the input output relations is not presented by a mapping (it is presented by a one-to-many relation). Each element (a,b) of the relation is then associated with a conditional probability P(b/a) of occurrence of b when a occurs. Probabilistic systems can be subdivided into :

i - Memoryless probabilistic systems. All output components are defined on the basis of instantaneous values of input components.

ii - Sequential probabilistic systems. At least one output component is not defined by the instantaneous values of input components.

In a sequential system the output (response) depends not only on the instantaneous input (stimulus) but also on the preceding inputs (stimuli). This means, however, that the required stimuli must be remembered by the system in the form of values of some internal quantities. Let us term them memory quantities, and the aggregate of their instantaneous values the internal state of the system.

The response of a deterministic system depends always uniquely on its internal state and on the stimulus. For stochastic systems the response depends only in probability on both the input (stimulus) and the internal state.

A system may be modeled using different abstractions. For example using graphs, deterministic or stochastic calculus, computer languages, etc. This raises the question about the equivalency of models and abstractions. From the view point of behavior, models are considered as equivalent when they produce similar sequences of outputs for similar sequences of inputs. Equivalency of abstractions may be related to the equivalence of their information measures (e.g. in Shannon's sense).

1.2 ENVIRONMENT.

Every system has its environment. With physical systems the environment is theoretically everything that is not included in the given system. However, since we confine ourselves mostly to a finite number of defined relations between the system and its environment, it is usually of advantage to restrict oneself to the substantial environment, i.e. to a limited set of elements which interest us in the environment. The same applies to abstract systems. The physical system and its environment act on each other - they interact. The manner of which a system influences its environment depends, in general, on the properties of the system itself as well as on the manner of which the environment acts on the system. Conversely, the same applies to the environment.

There is no "hard" boundary between the system and the environment. The environment is indeed another system which surrounds. The interaction process between the system and the environment can only continue when the environment defined by its behavior[2] and the system likewise form two abstract sets which neither include nor exclude each other. The intersection of the two sets represents the boundary between the system and the environment, see Fig.1. This represents the part of the system which is relevant to the environment and reversely the part of the environment which is relevant to the system : Relevance with regards to the environment's or the system's purposes or goals or means of their realization in work. So the boundary is representing the interaction context between the system and the environment. This interaction is maintained by both interdependence of purpose - complementarity and interaction which maintains separation and thus contradiction. Such tendancy both towards conflict and union are present in any real interacting process. The context is being metabolized as the system and its environment work at changing each other according to the dynamics of the interaction process and change into each other.

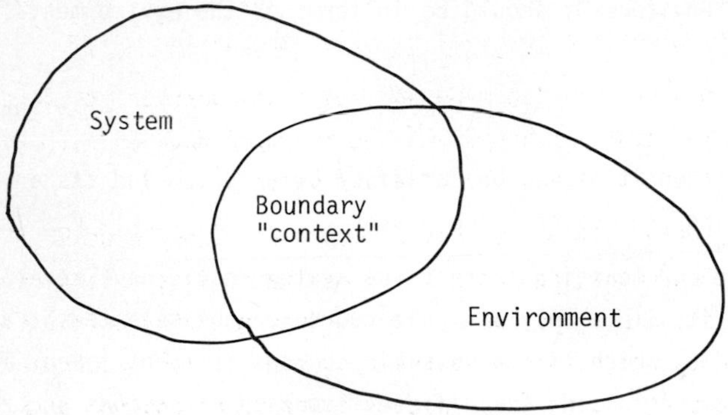

Fig.1 System-Environment Interaction.

1.3 CONTROL.

We shall for purposes of simplicity define control sequentially as first making a decision and then taking an action. Control is defined as aiming at a certain objective. Without objective there would be no decision : The word decision will be just meaningless. The objective can be considered as a function of a longer term decision and action determined by another system or system level (an aspect of environment). The decision and action of a system are directly related to the internal state and output of the system, respectively.

Taking a decision is equivalent to enhancing the order or organization of the system's state. If the system transfers part of its organization to the environment then a control process will be taking place. This amounts to information transfer between the system and the environment. Thus an action may be regarded as information transfer (energy and material are used in this transfer but this aspect will not be discussed in the presentation). A necessary condition enabling information to be received by the environment is that the system action be expressed by signals. However, every environment is capable of directly receiving some types of signal only (it is selective to the reception of signals), and it can received these signals only at given resolution level and during receptive intervals.

Information can be expressed by various signals and conversely, various meanings can be attributed to a given signal. Signals can be more or less ambiguous. In order to ensure that signals have meanings as information, there must exist a set of rules according to which a certain informational meaning is expressed is assigned to individual signals. This meaning determines the action or work that will be performed by using the information in relation to a purpose. A set of such rules determining the information value of a set of signals is called a code. Hence the system action to be interactive with the environment should be in terms of the environmental code.

1.4 LEARNING CONDITIONS.

Let us now consider the conditions to be satisfied by a system and its environment in order that a learning process can take place.

An obvious prerequisite for learning is that the system should have several courses of actions open to it. Since only a single course can be selected at any given time, it must be decided which of the possible courses is to be taken. This is equivalent to ordering or organizing the different courses of actions according to their preference in view of a certain goal linked to the environment response. The more the disorder of those courses of actions the more the need of learning. Entropy is defined as a measure of that disorder.

Let the set of actions be $Y = \{y_1, y_2, \ldots, y_n\}$. Define p_i as the probability of deciding for the course of action y_i $(i = 1, \ldots, n)$. Note that

$$0 \leq p_i \leq 1 \quad , \quad \sum_{i=1}^{n} p_i = 1 \tag{1}$$

The entropy H measuring the decision disorder is given by

$$H = -k \sum_{i=1}^{n} p_i \ln p_i \tag{2}$$

where k is some positive constant.

Then a prerequisite for learning is that the initial entropy H_o be greater than zero, i.e.

$$H_o > 0 \tag{3}$$

Besides the necessity for initial system's action disorder, it is also important for learning process to take place that the system be sensitive to the environment response. A receptive code is necessary. Obviously, if the system is insensitive or indifferent to the environment's response there will be no sense talking about learning for the environment would have no influence whatsoever on the system.

Thus the system's structure must be changing in accordance with the environment's response and in such a way that the system's entropy be decreasing with time passing, i.e.

$$\lim_{t \to \infty} H \rightarrow 0 \qquad (4)$$

1.5 LEARNING AND ENTROPY.

To understand the interrelationship between learning and entropy let us cite the following example. Suppose we give a multiple choice exam to a student. Even though he has to choose one of the alternatives, he is, except in a special case, not 100 percent sure that his choice is the correct one. In general, his degree of knowledge is better represented by the probability (or plausibility) distribution over the n alternatives. If his degree of learning is low, the probability will be distributed more or less evenly over the alternatives. As he learns more, however, the probability will be more and more concentrated on fewer and fewer alternatives. If one defines the entropy H as in eqn. (2) where p_i is the probability of the i-th alternative, the process of learning will be expressed by a decrease of entropy. It is true that the increase of confidence on a wrong alternative will be expressed by a decrease of entropy too, but we cannot deny that one aspect of the process of correct learning can be expressed by a decrease of the learning entropy. If one starts with a wrong belief and if one gradually shifts the weight from the wrong alternative to the correct alternative, the entropy in the first stage will increase, expressing the unlearning of a wrong belief, and in the second stage will decrease, expressing the learning of a correct belief, see Fig. 2. So unlearning becomes necessary for a successfuly succeeding phase of learning.

1.6 TYPES OF LEARNING SYSTEMS.

One important property of learning systems is their ability to demonstrate an improving performance in spite of lacking a priori information or even under conditions of initial indeterminacy.

Depending on the information input ot the system and the system - environment interaction it is possible to distinguish different types of learning processes :

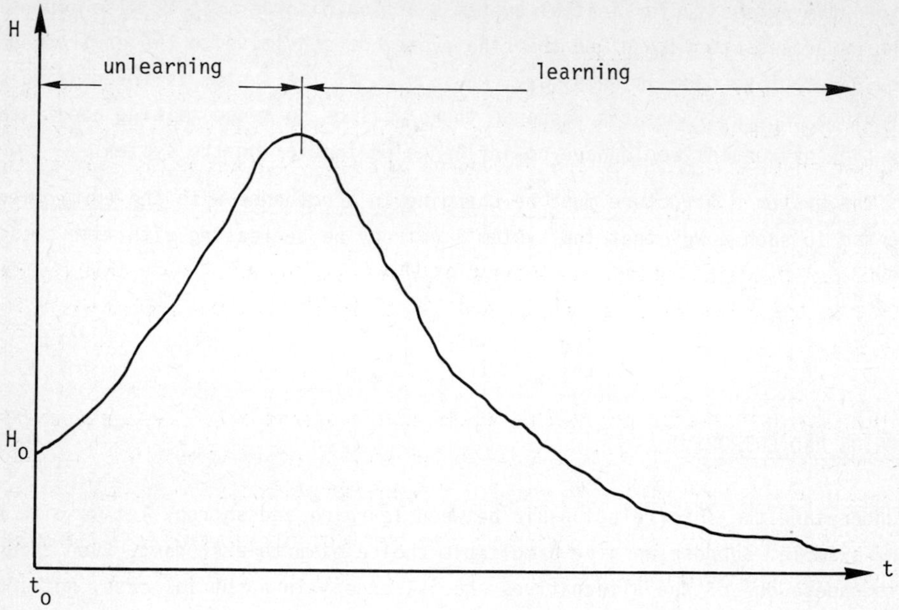

Fig.2. Evolution of learning system's entropy.

a. <u>Unsupervised Learning</u>.
(<u>Self-Leraning or Learning without Teacher</u>).

This is the case when the system does not receive any outside information except the ordinary signals from the environment. The system would then be learning by experimenting behavior. Such learning systems are usually called self-organizing systems, see Fig.3. The study of such systems finds its application for example, in problems of simulation of behavior, and automatic clustering of input data.

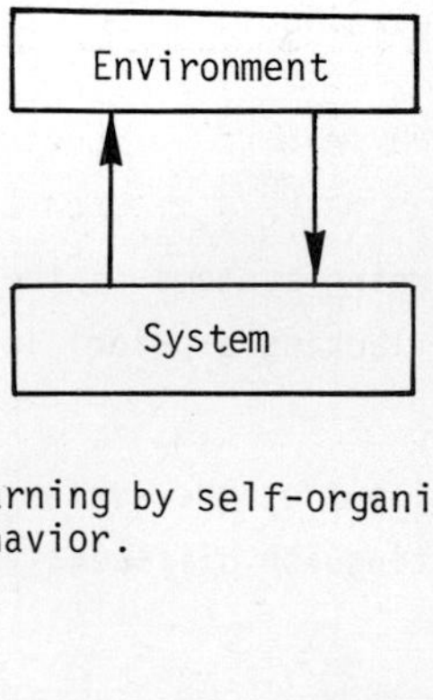

Fig.3. Learning by self-organization or experimenting behavior.

b. Supervised Learning.
(Training or Learning by a Teacher).

This is the case when the system receives additional information from the outside during the learning process. Here the teacher is the source of additional external information input to the system. Depending on what information the teacher inputs to the system undergoing training it is possible to distinguish two situations :

i. training by showing, the teacher inputs realizations of the output signal $\hat{y}$ corresponding to the given realizations of the input signal x to the system being trained, see Fig.4.

ii. training by assessment, the teacher observes the operation of the system being trained and inputs to it its appraisal z of the quality of its operation (in the simplest case the teacher gives a reward z = +1 or punishment z = -1), see Fig.4.

One may further classify learning by a teacher into two categories : learning by an ideal (or perfect) teacher, and learning by a real teacher (or teacher who makes mistakes).

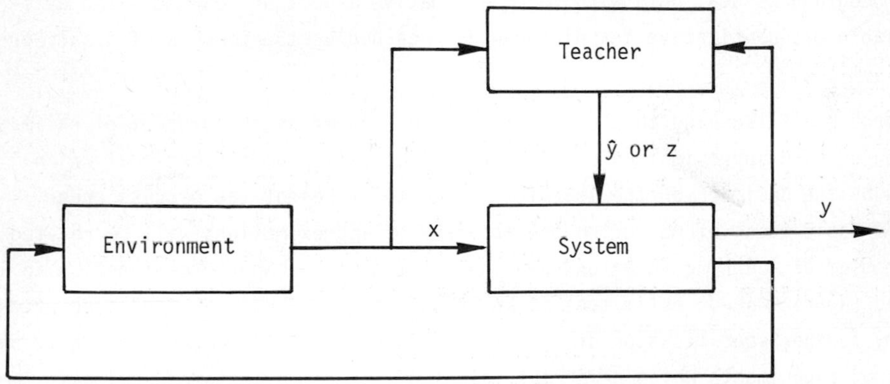

Fig.4. Learning by a Teacher, a)training by showing: the teacher inputs $\hat{y}$ to the system, b) training by assessment: the teacher inputs z to the system.

But there remains an important and yet unexplored type of learning where the interaction between the system and the teacher becomes active instead of passive. That is when the teacher like the system does not know the right action y for a situation x, the teacher and the system become both self-organizing systems learning from each other and experimenting their behaviors. There will in fact be no "specific" teacher ; the teacher in this case is just another system. Either system is the teacher for the other one. This case may be called cooperative learning in juxtaposition with the traditional (competitive) learning by a teacher. (It is valuable to note that that competitive learning requires cooperation in setting the rules within which the strategy will be developed, if the student has no code for observing this cooperative aspect then teacher and student will have difficulty in unlearning inferior learning techniques).

1.7 MATHEMATICAL MODELING.

Learning may be mathematically modelled as a hierarchy of experience problems, operating at different time-scales. The experience problems solved at the longer time span structure the problems to be at the shorter time spans, and vice versa. Experience problems of slow dynamics, corresponding to long time span, may be considered as qualitative learning. On the other side, experience problems of fast dynamics, corresponding to short time span, may be considered as quantitative learning. In this monograph we deal only with the quantitative aspect of learning. (An automaton capable of quantitative learning can be regarded as the simplex of a self-organizing system).

The problem of learning thus considered may be viewed as the problem of estimation or successive approximation of the unknown quantities of a functional which is chosen by the designer or the learning system to represent the process under study. The basic ideas of measuring the accuracy of approximations will be related to the problem of learning in an unknown environment, i.e., where the function to be learned (approximated, estimated) is known only by its form over the observation space. Any further specification of such a functional form can be performed only on the basis of experiments which offer the values of the approximated function in the domain of its definition. This implies that any desired solution which needs the knowledge of the approximated function is reached gradually by methods relying on experimentation and observation.

1.8 CONCLUSIONS.

The behavior of a system depends on its state of learning or information level which are measured by the organization of the system's output (action or response) corresponding to each input (or stimulus received from the environment). The higher the information level of the system the lower the entropy of the system. The optimal rule of the system's behavior, or the optimal relationship between its input and output, depends on the system's purpose, or goal. Learning is needed when a system does not know a priori the optimal rule of behavior, i.e. the initial entropy is greater than zero. Only by experimenting behavior or undergoing training by a teacher would the system then be able to learn about the optimal rule of behavior. Learning takes time. Throughout that time the system processes information and adapts its structure. If the system is learning successfuly then its entropy would decrease after a sufficiently large interval of time. The higher the learning rate the sharper would be the decrease in the system's entropy. The system might start learning with helding a wrong belief. So its entropy would be increasing instead of decreasing for some interval where the system would be unlearning. Finally we pointed out different types of leraning systems which can generally be classified as learning without teacher or learning with a teacher. We further classified the latter case into learning with ideal or real teacher. Learning with a teacher can further be classified as competive or cooperative.

COMMENTS.

1.1 Elaborate definitions on abstract systems and further details on general system modelling can be found in Klir and Valach[1], and Klir[2].

1.2 The comments on the system-environment-interaction cybernetic model are influenced by Brodey[3].

1.3 A good reference is Klir and Valach[1].

1.5 The example is quoted from Watanabe[4].

1.6 The models of unsupervised learning are important in behavioral science. Some models were introduced in the literature on behavioral psychology[5] and lately in engineering science[6]. A discussion on supervised learning, or training by assessment and by showing is given in Pugachev[7]. Some discussions on learning from a teacher who makes mistakes is given in Tebbe[8].

REFERENCES.

1. G.J.Klir, and M.Valach, Cybernetic Modelling. London : Iliffe books limited, 1956.
2. G.J.Klir, An approach to General Systems Theory. New York : Van Nostrad Reinhold, 1969.
3. W.Brodey, private discussions.
4. S.Watanabe, "Norbert Weiner and Cybernetical Concept of Time", IEEE Trans. on Syst., Man, and Cybern., May 1975, pp. 372-375.
5. R.R.Bush, and F.Mosteller, Stochastic Models for Learning. Wiley, 1958.
6. K.S.Narendra, and M.A.L.Thathachar, "Learning Automt- A Survey", IEEE Trans. Syst., Man, Cybern., vol. SMC-4, N°4, 1974,pp.323-334.
7. V.S.Pugachev, "Statistical Theory of Automatic Learning Systems", Izv. Akad. Nauk SSSR Eng. Cybern., N°6, 1967, pp. 24-40.
8. D.L.Tebbe, "On Learning from a Teacher who makes Mistakes", International Joint Conf. on Pattern Recognition Proceedings, Washington DC., Nov. 1973.

CHAPTER II

DECISION - Pattern RECOGNITION

César : C'est peu d'avoir vaincu puisqu'il faut vivre en doute.

Antoine : Mais s'en peut-il trouver un qui ne vous redoute ?

J. Grévins : La Mort de César.

2.1. PATTERN RECOGNITION PROBLEM.

The problem of pattern recognition is concerned with the analysis and the decision rules governing the identification or classification of observed situations, objects, or inputs in general.

For the purpose of recognition an input pattern $\underline{x}$ is characterized by a number of features constituting the elements of a feature vector $\underline{z}$. We shall for purpose of simplicity, decompose the pattern recognition problem into two problems :

i. - Feature extraction (characterization). Let the input pattern vector $\underline{x}$ lie in the pattern space Ω_x. The feature vector $\underline{z}$ lies in the feature space Ω_z and is constructed by effecting certain measurements or transformations on the input pattern.

While a pattern vector $\underline{x}$ might be infinite dimensional, the feature vector $\underline{z}$ as a rule is finite dimensional and usually of less dimension than $\underline{x}$. Hence the problem of characterization consists in finding an appropriate transformation T that maps the input pattern space into a feature space Ω_z such that $\underline{z}$ adequately characterizes the original $\underline{x}$ for purposes of classification, i.e. it provides enough information for discriminating the various patterns.

ii. - Abstraction and Generalization (Classification).

The abstraction problem is concerned with the decision rules for labeling or classifying feature measurements into pattern classes. The classification rules is such that the features in each class share more or less common properties.

Due to the distorted and noisy nature of feature measurements each pattern class could be characterized by certain statistical properties. Such properties may be fully-known, partially-known, or completely missing a priori. In the case of lacking a priori information it is required that the classifier undergoes training or to be built according to learning theorems.

2.2. FEATURE EXTRACTION.

The transformation $T : \Omega_x \to \Omega_z$ is usually characterized by a set of parameters called pattern features. A suitable set of pattern features should somehow reflect certain properties of the pattern classes.

The basic feature extraction problem can be classified into two general categories :

i - intraset feature extraction

ii - interset feature extraction.

Intraset feature extraction is concerned with those attributes which are common to each pattern class.

Interest feature extraction is concerned with those attributes characterizing the differences between or among pattern classes.

The intraset and interset features essentially pose conflicting extraction criteria. For intraset features the interest is to keep the distance (as a measure of dissimalirity) between the feature vectors belonging to the same class as close as possible to the distance between the corresponding pattern vectors. Alternatively stated the disorganization entropy between the sample vectors of the same class in the feature space is to be kept as close as possible to its value in the pattern space. This amounts to maximizing the entropy.

On the other hand for interset feature extraction the interest is to emphasize the differences between the patterns. This can be attained if some clustering of the same pattern samples is attained in the feature space. This amounts to contracting the distance between the same pattern samples in the feature space, thus enhancing the organization or minimizing the entropy.

2.3. KARHUNEN - LOEVE EXPANSION.

Assume there are K pattern classes $(K > 2)$ $w_1, w_2, \ldots w_k$. The pattern vector $\underline{x}$ is assumed to be N dimensional with probability density functions.

$$f(\underline{x}) = \sum_{k=1}^{K} \pi_k f_k(\underline{x}) \tag{1}$$

where π_k is the probability that a pattern belongs to class w_k, $f_k(\underline{x})$ is the conditional density of $\underline{x}$ for given w_k. We assume without loss of generality that $E(\underline{x}) = \underline{0}$, since a random vector with nonzero mean can be transformed into one with zero mean by translation, which is a linear operation. Then the covariance matrix $\underline{R}$ is the $N \times N$ matrix

$$\underline{R} = E\{\underline{x}\,\underline{x}^T\} = \sum_{k=1}^{K} \pi_k \, E_k \, \{\underline{x}\,\underline{x}^T\} \quad (2)$$

where E_k denotes the expectation over the pattern vectors of class w_k. The Karhunen-Loeve expansion is an expansion of the random vector $\underline{x}$ in terms of the eigenvectors of $\underline{R}$. Let λ_j ans $\underline{u}_j$ be the j-th eigenvalue and eigenvector of $\underline{R}$ i.e.

$$\underline{R}\,\underline{u}_j = \lambda_j\,\underline{u}_j \quad (3)$$

Since $\underline{R}$ is always symmetric and positive semi-definite it is easy to see that

$$\lambda_j \geqslant 0 \quad (4)$$

$$\underline{u}_j^T\,\underline{u}_\ell = 0 \qquad \text{if} \quad \lambda_j \neq \lambda_\ell \quad (5)$$

If $\underline{R}$ is further a full - rank matrix then there exists a set of N orthonormal eigen-vectors $\underline{u}_1$, $\underline{u}_2$, ..., $\underline{u}_N$ with eigenvalues $\lambda_1 \geqslant \lambda_2 \geqslant \ldots \geqslant \lambda_N \geqslant 0$.

The expansion in terms of eigenvectors,

$$\underline{x} = \sum_{j=1}^{N} c_j\,\underline{u}_j \quad ; \quad c_j = \underline{x}^T\,\underline{u}_j \quad (6)$$

is called the Karhunen - Loeve expansion. Note that c_j is a random variable due to the randomness of $\underline{x}$. Since we assume $E(\underline{x}) = \underline{0}$, $E(c_j) = 0$, and by (2), (3), and the orthonormality of $\underline{u}_j$,

$$E(c_j\,c_\ell) = E(\underline{u}_j^T\,\underline{x}\,\underline{x}^T\,\underline{u}_\ell) = \underline{u}_j^T\,\underline{R}\,\underline{u}_\ell = \lambda_j\,\delta_{j\ell} \quad (7)$$

In other words, the random variables c_j and c_ℓ are uncorrelated if $j \neq \ell$, and $E(c_j^2)$ equals the eigenvalue λ_j. This property of zero correlation is an important and unique property of the Kruhnen-Loeve expansion.

2.4. INTRASET FEATURE EXTRACTION.

The intraset feature extraction reflects the pattern properties common to the same class. Intraset feature extraction may be studied from various points of view. This extraction problem may be analyzed as an estimation problem, or considered as a problem of maximizing the population entropy (as noted before).

2.4.1. Estimation Problem.

Assume that the N-dimensional pattern vector $\underline{x}$ belongs to a multivariate population whose probability density $f(\underline{x})$ is gaussian with zero mean (i.e. $E(\underline{x}) = \underline{0}$)

and N x N covariance matrix $\underline{R}$.

Consider linear feature extraction, where the M - dimensional feature vector $\underline{z}$ (M < N) is given by the linear transformation

$$\underline{z} = T\,\underline{x} \tag{8}$$

where T is the matrix

$$T^T = (\underline{v}_1, \underline{v}_2, \ldots, \underline{v}_M) \tag{9}$$

where $\{\underline{v}_j\}$ is a set of orthonormal basis in Ω_x. Notice that the feature space Ω_z is a subspace of Ω_x whose basis is $\{\underline{v}_1, \ldots, \underline{v}_M\}$. If we expand $\underline{x}$ in terms of $(\underline{v}_j)$ we have

$$\underline{x} = \sum_{j=1}^{N} c_j\,\underline{v}_j \tag{10}$$

with the coefficients

$$c_j = \underline{x}^T\,\underline{v}_j \tag{11}$$

Note that c_j is a random variable due to the randomness of $\underline{x}$. The feature vector $\underline{z}$ becomes

$$\underline{z}^T = \underline{x}^T\,T^T = \sum_{j=1}^{N} c_j\,\underline{v}_j^T\,(\underline{v}_1, \ldots, \underline{v}_M) = (c_1, \ldots, c_M) \tag{12}$$

The last step is due to the orthonormality of $\underline{v}_j$. Thus the feature vector $\underline{z}$ consists of the first M coefficients.

The estimation problem consists in determining the matrix T, see eqn. (9) such that the error between the pattern vector $\underline{x}$ in Ω_x and its projection $\underline{z}$ in the feature space Ω_z be minimum. Mathematically stated it is required to determine the basis vectors $\underline{v}_1, \ldots, \underline{v}_M$ such that the error norm :

$$E||\underline{x}-\underline{z}||^2 = E\,\{(\underline{x}-\underline{z})^T(\underline{x}-\underline{z})\} = E\{(\sum_{j=M+1}^{N} c_j\underline{v}_j)^T(\sum_{k=M+1}^{N} c_k\underline{v}_k)\} \tag{13}$$

$$= E \sum_{j=M+1}^{N} c_j^2 = \sum_{j=M+1}^{N} E(c_j^2)$$

be minimum.

If one uses the Karhunen - Loeve expansion for the representation (10) then it follows from eqn. (7) that the required vectors $\underline{v}_1, \ldots, \underline{v}_M$ are given by the eigenvectors $\underline{u}_1, \ldots, \underline{u}_M$, see eqn.(3), corresponding to the M largest eigenvalues of the covariance matrix $\underline{R}$, see eqn. (2).

2.4.2. Entropy maximization.

Let Ω_x be the N - dimensional pattern space, and the feature space Ω_z be an M - dimensional subspace of Ω_x. The relationship between $\underline{x}$ and $\underline{z}$ may be expressed by $\underline{z} = T\,\underline{x}$ where T, an M x N matrix, is the linear feature extractor. The pattern vector $\underline{x}$ is distributed according to a continuous probability density function $f(\underline{x})$. Then the density function for $\underline{z}$, $f(\underline{z})$, is a marginal density of $f(\underline{x})$, and depends also on T. We define two entropies,

$$H(\underline{x}) = E_x -\{\ln f_x(\underline{x})\} = -\int_{\Omega_x} f_x(\underline{x}) \ln f_x(\underline{x})\, d\underline{x} \qquad (14)$$

$$H(\underline{z}) = E_z -\{\ln f_z(\underline{z})\} = -\int_{\Omega_z} f_z(\underline{z}) \ln f_z(\underline{z})\, d\underline{z}$$

We wish to find for feature extraction a matrix T that reduces the dimensionality to M and at the same time preserves as much information content as possible. This amounts to finding the M - space Ω_z that preserves the maximum entropy compared with other M - spaces. (Note that the entropy is a measure of the intraset dispersion). Let $f(\underline{x})$ be a Gaussian density with zero mean and covariance matrix R. The entropy then becomes

$$H(\underline{x}) = -\,E\,\{\ln f_x(\underline{x})\} = E\,\{\frac{N}{2} \ln 2\pi + \frac{1}{2} \ln |R| + \frac{1}{2}\,\underline{x}^T R^{-1} \underline{x}\} \qquad (15)$$

where $|R|$ is the determinant of R. Noting that

$$E\,\{\underline{x}^T R^{-1}\,\underline{x}\} = E\,\{\mathrm{tr}\,R^{-1}\,\underline{x}\,\underline{x}^T\} = \mathrm{tr}\,I = N \qquad (16)$$

we obtain

$$H(\underline{x}) = \frac{1}{2} \ln\ R + \frac{N}{2} \ln 2\,\pi + \frac{N}{2} \qquad (17)$$

Let $\underline{z} = T\,\underline{x}$ and T be an M x N matrix with orthonormal row vectors. Since the marginal density of a Gaussian distribution is Gaussian, we have

$$H(\underline{z}) = \frac{1}{2} \ln |R_z| + \frac{M}{2} \ln 2\pi + \frac{M}{2} \qquad (18)$$

where

$$R_z = T\,R\,T^T \qquad (19)$$

is the covariance matrix of $\underline{z}$. Since the determinant of a matrix is equal to the product of its eigenvalues, (18) may be written as

$$H(\underline{z}) = \frac{1}{2} \sum_{j=1}^{M} \ell n \; \phi_j + \frac{M}{2} \ell n \; 2\pi + \frac{M}{2} \quad (20)$$

with ϕ_j being the eigenvalues of the covariance matrix R_z. Hence we obtain the following result,

Theorem

Let $f(\underline{x})$ be a Gaussian density function with zero - mean and covariance matrix R. The optimum M x N linear feature extractor that maximizes H ($\underline{z}$) is

$$T^T = (\underline{u}_1, \underline{u}_2, \ldots, \underline{u}_M) \quad (21)$$

where $\underline{u}_1$, u_2, ..., $\underline{u}_M$ are the eigenvectors associated with the M largest eigenvalues $\lambda_1, \lambda_2, \ldots, \lambda_M$ in the Karhunen - Loeve expansion. The maximum entropy is

$$H(\underline{z}) = \frac{1}{2} \sum_{j=1}^{M} \ell n \; \lambda_j + \frac{M}{2} \ell n \; 2\pi + \frac{M}{2} \quad (22)$$

2.5. INTERSET FEATURE EXTRACTION.

So far we have discussed feature extraction without considering discrimination between different classes. Since pattern recognition is concerned with classification of patterns, an obvious criterion for feature extraction is the error probability. We would like to find an M - dimensional subspace of Ω_x such that the probability of classification errors is minimum compared with other M - subspaces. Unfortunately the error probability is generaly very difficult to calculate and it is practically impossible to use as a criterion for feature extraction.

Interset feature extraction is concerned with generating a set of features which tend to emphasize the dissimalirities between pattern classes. Kullback[21] has suggested that divergent information or divergence can provide an appropriate measure of the dissimilarities between two populations.

2.5.1. The Divergence.

Consider two classes of pattern w_1 and w_2 with probability density functions $f_1(\underline{x})$ and $f_2(\underline{x})$. From statistical decision theory see sec 2.7., the classification of a pattern $\underline{x}$ is based on the log. likelihood ratio,

$$\ell n \; \Lambda(\underline{x}) = \ell n \frac{f_2(\underline{x})}{f_1(\underline{x})} \quad (23)$$

If $\ln \Lambda(\underline{x})$ is greater than a certain threshold value, $\underline{x}$ is classified as belonging to w_2 ; otherwise to w_1.

Therefore, we define

$$J_1(\underline{x}) = E_1\{ \ln \frac{f_1(\underline{x})}{f_2(\underline{x})} \} = \int_{\Omega_x} f_1(\underline{x}) \ln \frac{f_1(\underline{x})}{f_2(\underline{x})} d\underline{x}$$
$$J_2(\underline{x}) = E_2\{ \ln \frac{f_2(\underline{x})}{f_1(\underline{x})} \} = \int_{\Omega_x} f_2(\underline{x}) \ln \frac{f_2(\underline{x})}{f_1(\underline{x})} d\underline{x} \qquad (24)$$

where $E_1\{.\}$ and $E_2\{.\}$ indicate the expectation over the densities $f_1(\underline{x})$ and $f_2(\underline{x})$ respectively. $J_1(\underline{x})$ may be interpreted as the average information for discrimination in favor of w_1 against w_2, and $J_2(\underline{x})$ may be interpreted in a similar manner. The divergence is defined as

$$J(\underline{x}) = J_1(\underline{x}) + J_2(\underline{x}) \qquad (25)$$

and is therefore a measure of information for discrimination of the two classes.

The measure (25) stated for the two-class case can be converted to the K - class case by optimizing the sum of all pairwise measures of quality or by maximizing the minimum of pairwise measure of quality.

When $\underline{x}$ is replaced by $\underline{z}$ and the densities f_x are replaced by the transformed densities f_z the criterion (25) can also measure the overlap of the class-conditional probability densities of the transformed samples.

Fig.1 illustrates two possible one dimensional distributions of samples resulting from two transformations applied to the same distribution in Ω_x. The transformation which results in the least overlap of probability densities will yield the space with the least expected classification error with respect to the Bayes optimal decision rule. The measure (25) is optimized when there is no overlap and take its worst value when the class densities are identical (maximum overlap).

In defining (24) we have used the convention that $f_1(\underline{x})/f_2(\underline{x}) = 0$ if $f_1(\underline{x}) = f_2(\underline{x}) = 0$ and $0 \ln \infty = 0$. It is interesting to note that when the two classes are separable, i.e. $f_2(\underline{x}) = 0$ if $f_1(\underline{x}) > 0$ and vice versa, the patterns may be classified without error and $J(\underline{x}) = \infty$. On the other hand, when $f_1(\underline{x}) = f_2(\underline{x})$ for almost all $\underline{x}$, the two classes are indistinguishable and $J(\underline{x}) = 0$.

2.5.2. Feature extraction.

Let us now discuss the application of the divergence criterion to the following simple examples.

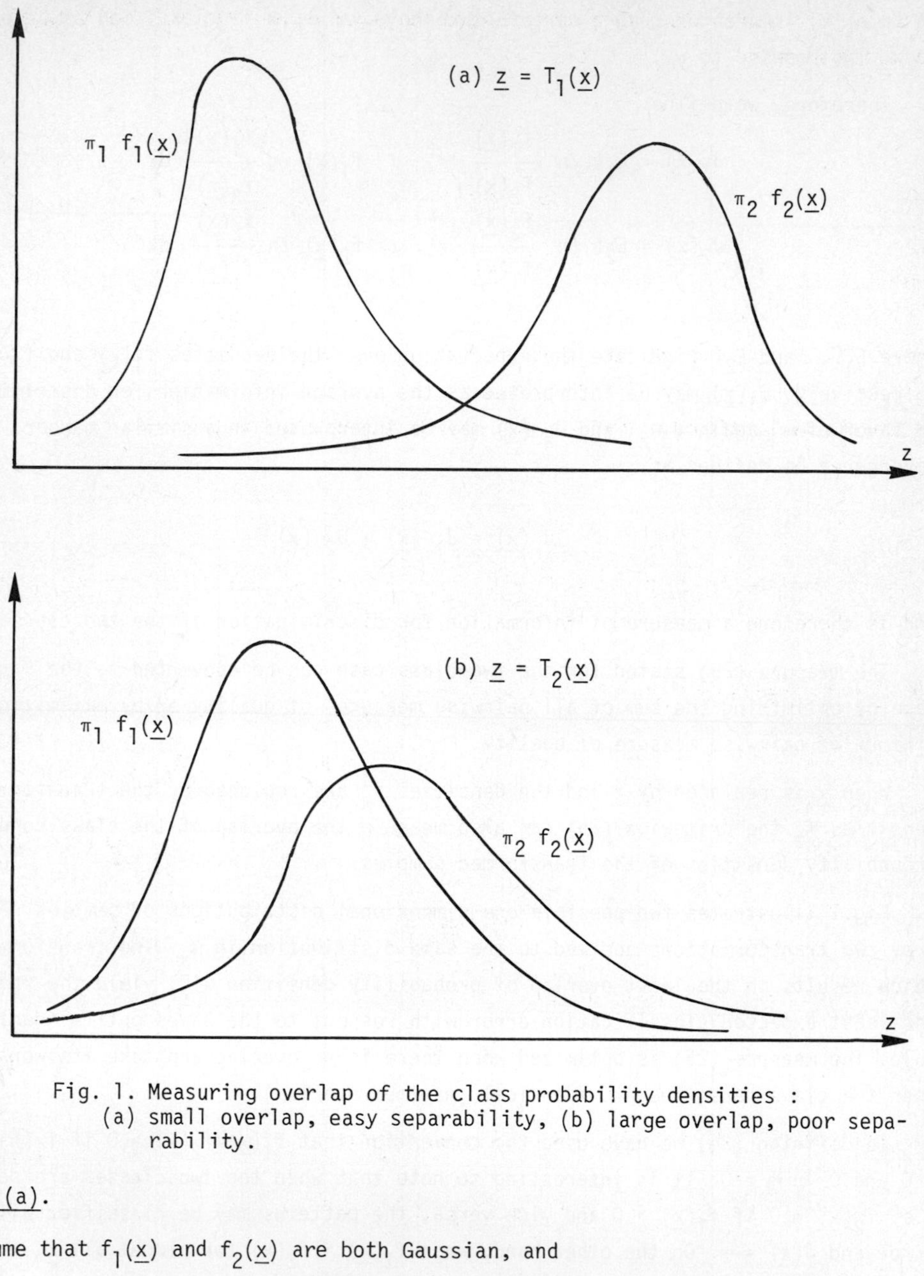

Fig. 1. Measuring overlap of the class probability densities : (a) small overlap, easy separability, (b) large overlap, poor separability.

Example (a).

Assume that $f_1(\underline{x})$ and $f_2(\underline{x})$ are both Gaussian, and

$$E_1(\underline{x}) = \underline{0} \quad , \quad E_2(\underline{x}) = \underline{m}$$
$$E_1(\underline{x}\,\underline{x}^T) = R_1, \quad E_2(\underline{x}\,\underline{x}^T) = R_2 \tag{26}$$

For a linear feature extractor T, the marginal densities, $f_1(\underline{z})$ and $f_2(\underline{z})$, are Gaussian with

$$E_1(\underline{z}) = \underline{0} \qquad E_2(\underline{z}) = \underline{m}_z$$
$$E_1(\underline{z}\,\underline{z}^T) = R_{z1} \qquad E_2(\underline{z}\,\underline{z}^T) = R_{z2} \tag{27}$$

where

$$\underline{m}_z = T\,\underline{m} \;,\quad R_{z1} = T\,R_1\,T^T \;,\quad R_{z2} = T\,R_2\,T^T \tag{28}$$

We obtain by straightforward calculation the divergence measure

(29)

$J(\underline{x})$ is similar to $J(\underline{z})$ in (29) with $\underline{m}_z$, R_{z1} and R_{z2} substituted by $\underline{m}$, R_1 and R_2. Let us consider two special cases.

(1) Equal covariance case. In this case

$$R_1 = R_2 = R \;,\quad R_{z1} = R_{z2} = R_z$$

and obviously

$$J(\underline{x}) = \underline{m}^T R^{-1}\,\underline{m}$$

If we select a 1 x N linear feature extractor,

$$T = \underline{m}^T R^{-1} \tag{30}$$

and substitute (28) and (30) into (29), we obtain

$$J(z) = \underline{m}^T R^{-1}\,\underline{m}\,(\underline{m}^T R^{-1} R\, R^{-1}\,\underline{m})^{-1}\,\underline{m}^T R^{-1}\,\underline{m} = J(\underline{x}) \tag{31}$$

The result suggests that the other directions do not contribute to the discrimination of the two classes. In other words the optimum classification is based on the statistic $\underline{m}^T R^{-1}\,\underline{x}$.

b) Equal means. In this case the mean of the second-class is also zero, $\underline{m} = \underline{0}$. If both R_1 and R_2 are positive definite then there exists a real and non-singular N x N matrix U,

$$U^T = (\underline{u}_1, \ldots, \underline{u}_N) \text{ such that}$$

$$U R_1 U^T = \Lambda \quad , \quad U R_2 U^T = I \tag{32}$$

where Λ is a diagonal matrix with real and positive elements $\lambda_1, \lambda_2 \ldots, \lambda_N$ and I is the identity matrix. In fact, the row vectors of U are the solutions of the equation,

$$R_1 \underline{u}_j = \lambda_j R_2 \underline{u}_j \tag{33}$$

It is noted that (32) implies a weighted orthonormality condition

$$\underline{u}_j^T R_2 \underline{u}_j = 1 \quad , \quad \underline{u}_j^T R_2 \underline{u}_\ell = 0 \quad j \neq \ell \tag{34}$$

Since U is non singular and J ($\underline{x}$) is invariant under non singular transformations, we may use (32) to calculate J (z) with R_{z1} and R_{z2} substituted by Λ and I. Thus

$$J(z) = 1/2 \text{ tr } (\Lambda - I)(I - \Lambda^{-1}) = 1/2 \sum_{j=1}^{N} (\lambda_j + \frac{1}{\lambda_j} - 2) \tag{35}$$

Eqn. (35) indicates that we should choose a feature extractor

$$T^T = (\underline{u}_1 , \underline{u}_2, \ldots, \underline{u}_M) \tag{36}$$

where $\underline{u}_j$ is associated with λ_j which is ordered according to

$$\lambda_1 + \frac{1}{\lambda_1} > \lambda_2 + \frac{1}{\lambda_2} > \ldots > \lambda_N + \frac{1}{\lambda_N} \tag{37}$$

The resulting value of J($\underline{z}$) is

$$J(z) = 1/2 \sum_{j=1}^{M} (\lambda_j + \frac{1}{\lambda_j} - 2) \tag{38}$$

Note that the row vectors of T are orthonormal in the sense of (34) instead of $\underline{u}_j^T \underline{u}_\ell = 0$, which is a property of the optimal T considered in the previous sections.

2.6. OPTIMAL CLASSIFICATION.

Algorithms of optimal classification can be obtained if there is enough a priori statistical data about each pattern class. The analytical tool to obtain such classification rules is the statiscal decision theory.

Let the pattern classes be $w_1, w_2, \ldots, w_K$. For each pattern class w_j, $j=1,..,K$, assume that the conditional multivariate (M - dimensional) probability density of the feature vector $\underline{z}$ (M - dimensional), $f_j(\underline{z})$ as well as the probabilities π_j of occurences of w_j ($i = 1, \ldots, K$) are known.

The problem of classification consists in partitioning the feature space Ω_z into K subspaces $\Gamma_1, \Gamma_2, \ldots, \Gamma_K$, see Fig.2, such that if $\underline{z} \in \Gamma_i$ we classify the pattern to the class w_i.

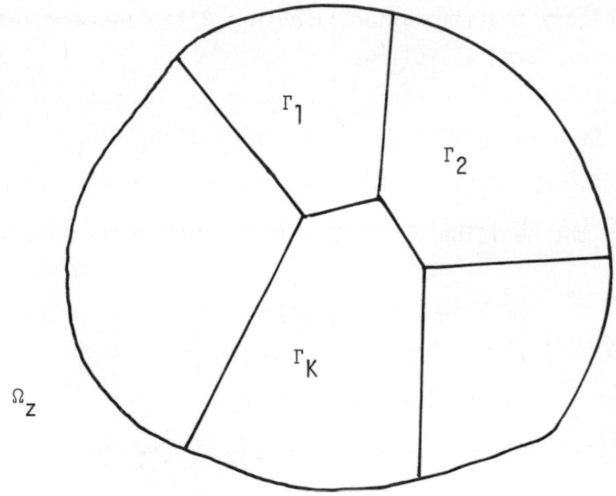

Fig. 2. Partitioning of the feature space

In order to describe the concept of "best" partition, we introduce the loss function

$$F(w_i, y) \quad , \quad i = 1, \ldots, K \tag{39}$$

where y is the classification decision, such that if $\underline{z} \in \Gamma_j$ then $y = y_j$, $j = 1,.., K$. Hence $F(w_i, y_j)$ denotes the loss to be incurred when a feature from the i th pattern is classified into the j-th class.

The conditional loss for $\underline{z} \sim w_i$ is

$$r(w_i, y) = \int_{\Omega_z} F(w_i, y) f_i(\underline{z})\, d\underline{z} \tag{40}$$

For a given set of a priori probabilities $\underline{\pi} = (\pi_1, \ldots, \pi_K)^T$, the average loss is

$$R(\underline{\pi}, y) = \sum_{i=1}^{K} \pi_i \, r(w_i, y) \tag{41}$$

substituting (40) into (41) we get

$$R(\underline{\pi}, y) = \int_{\Omega_Z} \sum_{i=1}^{K} F(w_i, y) \, f_i(\underline{z}) \, \pi_i \, d\underline{z} \tag{42}$$

The problem is to find $y \varepsilon \{y_1, \ldots\ldots, y_K\}$ as a function of $\underline{z}$, such that the average loss is minimized.

In the case of binary classification, i.e. K = 2 the average loss function given by eqn. (42) can be rewritten as follows,

$$R(\underline{\pi}, y) = \int_{\Omega_Z} F(w_1, y) \, f_1(\underline{z}) \, \pi_1 + F(w_2, y) \, f_2(\underline{z}) \pi_2 \, d\underline{z} \tag{43}$$

By definition of the decision variable y it follows that

$$\int_{\Omega_Z} F(w_i, y_j) \, f_i(\underline{z}) \, \pi_i \, d\underline{z} = \int_{\Gamma_Z} F(w_i, y_j) \, f_i(\underline{z}) \, \pi_i \, d\underline{z} \qquad (i, j = 1, 2) \tag{44}$$

Let us define the <u>conditional error probability of the first kind.</u>

$$\alpha = \int_{\Gamma_2} f_1(\underline{z}) \, d\underline{z} \tag{45}$$

corresponding to classifying an observation from w_1 into w_2. Similarly, define the <u>conditional error probability of the second kind</u>

$$\beta = \int_{\Gamma_1} f_2(\underline{z}) \, d\underline{z} \tag{46}$$

If we consider the loss function

$$\begin{aligned} F(w_1, y_1) &= \nu_{11} \quad , \quad F(w_2, y_1) = \nu_{21} \\ F(w_1, y_2) &= \nu_{12} \quad , \quad F(w_2, y_2) = \nu_{22} \end{aligned} \tag{47}$$

$$\nu_{11} < \nu_{12} \qquad\qquad \nu_{22} < \nu_{21}$$

then the average loss (43) can be expressed as

$$R(\underline{\pi}, y) = (\nu_{11}(1-\alpha) + \nu_{12}\alpha)\pi_1 + \\ + (\nu_{21}\beta + \nu_{22}(1-\beta))\pi_2 \tag{48}$$

2.7. STATISTICAL DECISION ALGORITHMS.

2.7.1. Bayes Approach.

In such case the average loss is assumed to have the general form (42). Minimization of that function with respect to y is equivalent to minimizing

$$F(w_1, y) f_1(\underline{z}) \pi_1 + F(w_2, y) f_2(\underline{z}) \pi_2 \tag{49}$$

for any observation or feature vector $\underline{z}$. Since the decision y takes only two values y_1 and y_2 say + 1 and - 1, respectively, then minimization of (49) can be obtained by simply comparing the corresponding values.

$$\begin{aligned} &\nu_{11} f_1(\underline{z}) \pi_1 + \nu_{21} f_2(\underline{z}) \pi_2 \quad \text{for } y=y_1 \\ &\nu_{12} f_2(\underline{z}) \pi_1 + \nu_{22} f_2(\underline{z}) \pi_2 \quad \text{for } y=y_2 \end{aligned} \tag{50}$$

Hence to minimize (49) we conclude the decision rule,

$$\begin{aligned} &y = +1 \text{ if } \nu_{11} f_1(\underline{z}) \pi_1 + \nu_{21} f_2(\underline{z}) \pi_2 < \nu_{12} f_1(\underline{z}) \pi_1 + \nu_{22} f_2(\underline{z}) \pi_2 \\ &y = -1 \text{ if } \nu_{11} f_1(\underline{z}) \pi_1 + \nu_{21} f_2(\underline{z}) \pi_2 > \nu_{12} f_1(\underline{z}) \pi_1 + \nu_{22} f_2(\underline{z}) \pi_2 \end{aligned} \tag{51}$$

The decision rule (51), called <u>Bayes rule</u>, can be rewritten in the form,

$$\begin{aligned} &y = +1 \quad \text{if} \quad \lambda(\underline{z}) > h \\ &y = -1 \quad \text{if} \quad \lambda(\underline{z}) < h \end{aligned} \tag{52}$$

where $\lambda(\underline{z})$ denotes the <u>likehood ration</u>

$$\lambda(\underline{z}) = \frac{f_1(\underline{z})}{f_2(\underline{z})} \tag{53}$$

and h is the threshold

$$h = \frac{\nu_{21} - \nu_{22}}{\nu_{12} - \nu_{11}} \cdot \frac{\pi_2}{\pi_1} \quad (54)$$

The decision rule (52) can be implemented as shown in Fig. 3.

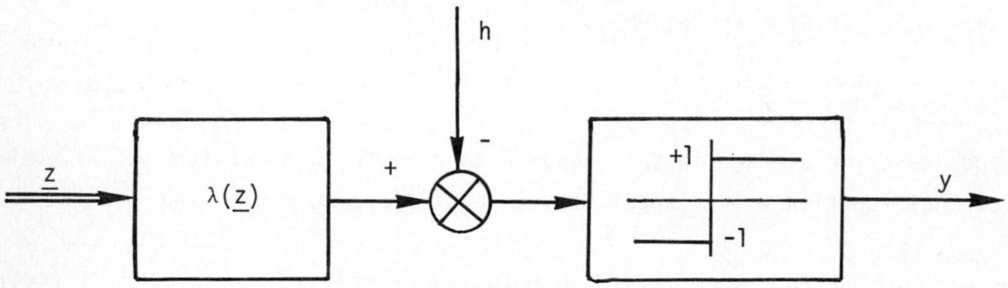

Fig.3. Binary statistical classification.

The particular choice of the ν's

$$\nu_{11} = \nu_{22} = 0 \quad , \quad \nu_{12} = \nu_{21} = 1 \quad (55)$$

leads to the classifier minimizing the criterion

$$R = \alpha\pi_1 + \beta\pi_2 \quad (56)$$

which amounts to the total probability of error. The decision rule can then be expressed by (52) with the threshold value

$$h = \pi_2 / \pi_1 \quad (57)$$

This rule is called the Siegert - Kotelnikov.

Another particular choice of the ν's is

$$\nu_{11} = \nu_{22} = 0 \quad , \quad \nu_{12} = 1 \quad , \quad \nu_{21} = \mu \quad (58)$$

This corresponds to the mixed-decision rule where

$$R = \alpha\pi_1 + \mu\beta\pi_2 \quad , \quad \mu > 0 \tag{59}$$

and

$$h = \mu\ \pi_2 \ /\ \pi_1 \tag{60}$$

The constant μ in (59) is indeed reminiscent to the Lagrange multiplier. Considering $\mu\ \frac{\pi_2}{\pi_1}$ as Lagrange multiplier interprets criterion (59) as minimizing the conditional error probability of the first kind α subject to a constant conditional error probability of the second kind β, i.e.

$$\beta = \int_{\Gamma_1} f_2(\underline{z})\ d\underline{z} = A = \text{const.} \tag{61}$$

This actually corresponds to the <u>Neyman-Pearson</u> rule for which the threshold h is given by (60) where μ is determined by solving the implicit relation,

$$\int_h^\infty f_2\ (\underline{z})\ d\lambda\ (\underline{z}) = A \quad , \quad h = \mu\ \frac{\pi_2}{\pi_1} \tag{62}$$

2.7.2. <u>Min - Max Approach</u>.

The Bayes approach necessitates knowledge of the a priori probabilities π_1 and π_2. If that knowledge is lacking or such probabilities change under different environmental conditions then one possible approach is to optimize the worst case. The min - max approach can be interpreted as a "game with nature" where "nature" choses the a priori probabilities that maximize the average risk. Let us consider.

$$\nu_{11} = \nu_{22} = 0 \tag{63}$$

The optimal Bayes rule is determined by minimizing the average loss,

$$R\ (\underline{\pi},\ y) = \nu_{12}\ \hat{\pi}_1\ \alpha + \nu_{21}\ (1 - \hat{\pi}_1)\ \beta \tag{64}$$

where $\hat{\pi}_1$ is an estimate of the a priori probability π_1. Minimizing (64), in its equivalent form (49) with respect to y yields the same rule as (52) with the threshold

$$h = \frac{\nu_{21}}{\nu_{12}}\ \frac{(1 - \hat{\pi}_1)}{\hat{\pi}_1} \tag{65}$$

The conditional error probabilities of the first - and second - kind, α, β, respectively (eqs. (45), (46)) will be functions of the estimated a priori probability $\hat{\pi}_1$. That is, $\alpha = \alpha(\hat{\pi}_1)$, $\beta = \beta(\hat{\pi}_1)$. If the actual value of π_1 is π_1^o then the adoption of the decision rule (52) with the threshold (65) will yield the following deviation between the actual average loss and its estimated optimal value,

$$\Delta R\ (\hat{\pi}_1, \pi_1^{o}) = (\nu_{12}\ \alpha(\hat{\pi}_1) - \nu_{21}\ \beta(\hat{\pi}_1))(\pi_1^{o} - \hat{\pi}_1)$$

The approach of min - max consists of choosing $\hat{\pi}_1$ which minimizes the maximum deviation. Hence $\hat{\pi}_1$ is determined by the condition,

$$\nu_{12}\ \alpha(\hat{\pi}_1) - \nu_{21}\ \beta(\hat{\pi}_1) = 0 \tag{66}$$

Equations (65) and (66) completely specify the min-max rule computational difficulties however can arise when solving equation (65).

2.7.3. Maximum A posteriori Probability Rule.

In the case of complete a priori information an intuitively appealing decision rule is the maximum a posteriori.

According to Bayes formula, a posteriori probabilities that an observed situation $\underline{z}$ belongs to classes w_1 or w_2 are equal respectively to

$$Pr\ (w_1\ /\ \underline{z}) = f_1\ (\underline{z})\ \pi_1\ /\ f\ (\underline{z}) \tag{67}$$

and

$$Pr\ (w_2\ /\ \underline{z}) = f_2\ (\underline{z})\ \pi_2\ /\ f\ (\underline{z}) \tag{68}$$

where

$$f\ (\underline{z}) = f_1\ (\underline{z})\ \pi_1 + f_2\ (\underline{z})\ \pi_2 \tag{69}$$

is the mixture probability density.

The observation $\underline{z}$ is classified into Γ_1 or Γ_2 depending on whether the a posteriori probability with respect to w_1 is greater than that with respect to w_2, or vice versa, respectively. According to eqs. (67) and (68) it follows immediately the decision rule (52) with the threshold.

$$h = \pi_2\ /\ \pi_1 \tag{70}$$

In the case of equal a priori probabilities, i.e.

$$\pi_1 = \pi_2 = 1/2 \tag{71}$$

the decision rule is called the maximum likehood.

2.8. SEQUENTIAL METHODS.

The algorithms presented so far are based on a fixed size M of feature measurements, or dimension of feature vector.

If the cost of taking feature measurements is to be considered or if the feature extracted from input patterns are sequential in nature then sequential classification methods are to be used.

In sequential methods feature measurements are processed sequentially on successive steps. At each step a decision is made either to further extract features or to terminate the sequential process (i.e. make the classification decision). The continuation or termination of the sequential process depends on a trade-off between the error (misrecognition) and the number of features to be measured. The process is terminated when a sufficient or desirable accuracy of classification has been achieved.

2.8.1. Wald's Sequential Probability Ratio Test (SPRT).

Suppose that a random variable $\underline{z}$ has the conditional probability density functions $f_i(\underline{z})$ for the pattern classes w_i, $i = 1, 2$.

The problem is to test the hypothesis $H_1 : z \sim w_1$ against the hypothesis $H_2 : z \sim w_2$. The test constructed decides in favor of either w_1 or w_2 on the basis of observations $z_1, z_2, \ldots$ The components $z_1, z_2, \ldots$ of the vector $\underline{z}$ are assumed to be independent and identically distributed random variables. Suppose that if w_1 is true we wish to decide for w_1 with probability at least $(1 - \alpha)$ while if w_2 is true we wish to decide for w_2 with probability at least $(1 - \beta)$.

Let us introduce the variables

$$\zeta_i = \ln \frac{f_1(z_i)}{f_2(z_i)} \quad , \quad i = 1, 2, \ldots \tag{72}$$

so that the likelihood ratio λ_n, corresponding to n observations, can be written thus

$$\lambda_n = \prod_{i=1}^{n} \frac{f_1(z_i)}{f_2(z_i)} = \frac{f_1(\underline{z})}{f_2(\underline{z})} \tag{73}$$

Wald's SPRT procedure is as follows : Continue taking observations as long as

$$B < \lambda_n < A \tag{74}$$

Stop taking observations and decide to accept the hypothesis H_1 as soon as

$$\lambda_n \geqslant A \tag{75}$$

and stop taking observations and decide to accept the hypothesis H_2 as soon as

$$\lambda_n \leqslant B \tag{76}$$

The constants A and B are called the upper and lower stopping boundaries respectively. They can be chosen to obtain approximately the probabilities of error α and β prescribed.

Suppose that at the n-th stage of the measuring process it is found that

$$\lambda_n = A \tag{77}$$

leading to the terminal decision of accepting H_1. From (73) and (77),

$$f_1\ (\underline{z}) = A\ f_2\ (\underline{z}) \tag{78}$$

which is equivalent to

$$\int_{\Gamma_1} f_1\ (\underline{z})\ d\underline{z} = A \int_{\Gamma_1} f_2\ (\underline{z})\ d\underline{z} \tag{79}$$

by the definitions of α and β, (79) reduces to

$$(1 - \alpha) = A\beta \tag{80}$$

Similarly, when

$$\lambda_n = B \tag{81}$$

then

$$\alpha = B\ (1 - \beta) \tag{82}$$

Solving (80) and (82), we obtain

$$A = (1 - \alpha) / \beta \tag{83}$$

$$B = \alpha / (1 - \beta) \tag{84}$$

It is noted that the choice of stopping boundaries A and B results in error probabilities α and β if continuous observations are made and the exact equality (77) and (81) can be respected.

From (77) and (81), again by neglecting the excess over the boundaries, we have

$$L_n = \ell n\ \lambda_n = \ell n\ A \text{ with probability } \beta \text{ when } H_2 \text{ is true}$$
$$L_n = \ell n\ B \text{ with probability } (1 - \beta) \text{ when } H_2 \text{ is true}$$
$$L_n = \ell n\ A \text{ with probability } (1 - \alpha) \text{ when } H_1 \text{ is true}$$
$$L_n = \ell n\ B \text{ with probability } \alpha \text{ when } H_1 \text{ is true}$$

Let $E_i\ (L_n)$ be the conditional expectation of L_n when H_i is true, then it follows directly that

$$E_1\ (L_n) = (1 - \alpha)\ \ell n\ A + \alpha \ell n\ B \tag{85}$$

$$E_2\ (L_n) = \beta \ell n\ A + (1 - \beta)\ \ell n\ B \tag{86}$$

Define

$$\eta_i = 1 \quad \text{if no decision is made up to the } (i - 1) \text{ the stage}$$
$$= 0 \quad \text{if a decision is made at an earlier stage}$$

Then η_i is clearly a function of $z_1, z_2, \ldots, z_{i-1}$ only and is independent of z_i and hence independent of $\zeta_i = \zeta_i\ (z_i)$.

$$L_n = \sum_{i=1}^{N} \zeta_i = \zeta_1\ \eta_1 + \zeta_2\ \eta_2 + \ldots. + \zeta_n\ \eta_n \tag{87}$$

Taking expectations

$$\begin{aligned} E\ (L_n) &= (\sum_{i=1}^{\infty} \zeta_i\ \eta_i) \\ &= E\ (\zeta) \sum_{i=1}^{\infty} E\ (\eta_i) \\ &= E\ (\zeta) \sum_{i=1}^{\infty} Pr\ (n \geqslant i) \qquad (88) \\ &= E\ (\zeta)\ E\ (n) \end{aligned}$$

Therefore, from (85) the average number of observations when H_1 is true can be expressed as

$$E_1^{(w)}(n) = \frac{(1-\alpha)\,\ell n\, A + \alpha\,\ell n\, B}{E_1\,(\zeta)} \tag{89}$$

Similarly, from (86),

$$E_2^{(w)}(n) = \frac{\beta\,\ell n\, A + (1-\beta)\,\ell n\, B}{E_2\,(\zeta)} \tag{90}$$

The superscript (w) is used to designate Wald's test.

2.8.2. Finite Automata.

As mentioned in section 2.8.1. the equalities (83) and (84) hold only if continuous observations are made so that the exact equality of (77) and (81) can be obtained.

A discrete from of Wald's test naturally needs a longer termination time but is simpler to realize. A device for its realization may be considered as a finite automaton with linear tactic, which may be described as follows.

The automaton has (s + 1) states numbered from 0 to s (Fig.4) and is characterized with the following system of transition : if the automaton is in the state j and as a result of an experiment $\zeta > a$ is obtained, where

$$\zeta = \ell n\,\frac{f_1\,(\underline{z})}{f_2\,(\underline{z})} \tag{91}$$

and "a" is some threshold, then the automaton passes to the state (j + 1) ; if $\zeta < b$, then the automaton passes to the state (j -1) ; if $b < \zeta < a$ then the automaton remains in state j. Motion begins from the state with index i. The states 0 and s are terminal : attainement by the automaton of state 0 leads to output of a decision in favor of hypothesis $H_2 : \underline{z} \sim w_2$; attainement of the state with index s, to a decision in favor of hypothesis $H_1 : \underline{z} \sim w_1$.

We shall consider the symmetric case, where the thresholds a and b are chosen so that $P\,(\zeta > a|\,H_1) = P\,(\zeta < b|\,H_2) = p$, $P(\zeta > a|\,H_2) = P(\zeta < b|\,H_1) = q$ and $r = 1-p-q = P(b < \zeta < a)$.

Let us define

$$L_j \triangleq \begin{cases} \text{Probability of the automaton attaining} \\ \text{the state s if it begins its motion} \\ \text{from the state j} \end{cases} \tag{92}$$

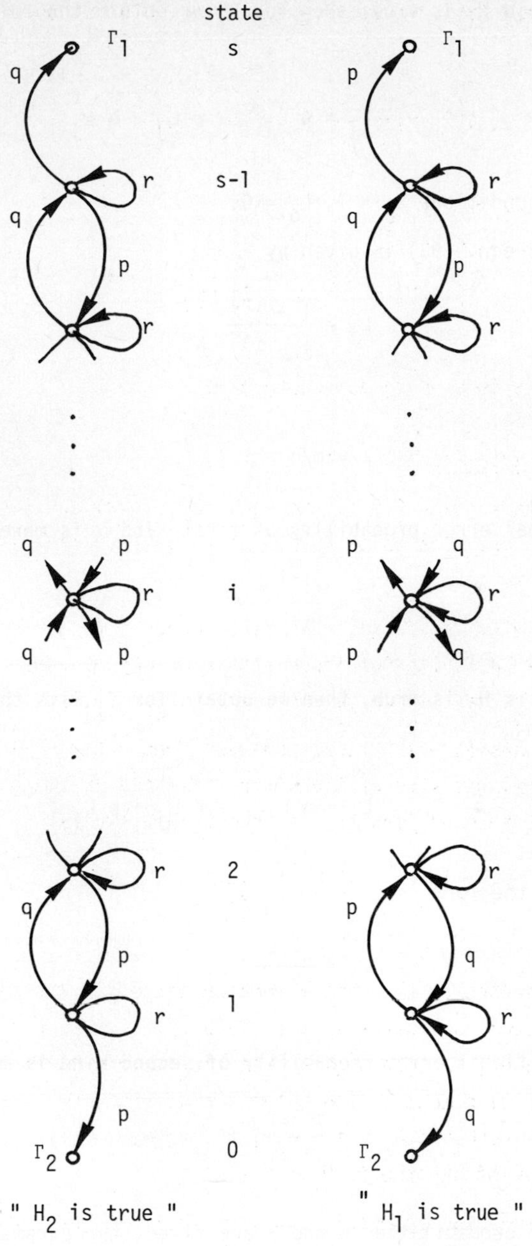

Fig. 4. Automaton with linear tactic.

If the hypothesis H_1 is true, then for L_j we obtain the following finite - difference equation :

$$L_j = p\, L_{j+1} + r\, L_j + q\, L_{j-1} \tag{93}$$

with the boundary conditions $L_s = 1$, $L_o = 0$.

The solution of eqn. (93) is given by

$$L_j = \frac{1 - \lambda^{-j}}{1 - \lambda^{-s}} \tag{94}$$

where

$$\lambda = p/q > 1 \tag{95}$$

Since the conditional error probability of first kind α is merely $1 - L_i$, we have

$$\alpha = \frac{\lambda^{s-i} - 1}{\lambda^{s} - 1} \tag{96}$$

If the hypothesis H_2 is true, then we obtain for L_j with the same boundary conditions, the equation

$$L_j = q\, L_{j+1} + r\, L_j + p\, L_{j-1} \tag{97}$$

whose solution has the form

$$L_j = \frac{\lambda^{j} - 1}{\lambda^{s} - 1} \tag{98}$$

Since the conditional error probability of second kind is merely L_i, we have

$$\beta = \frac{\lambda^{i} - 1}{\lambda^{s} - 1} \tag{99}$$

Hence, if the error probabilities α and β are fixed, the parameters s and i of the automaton will be given by

$$i = \ell n\, \left(\frac{1 - \beta}{\alpha}\right) / \ell n\, \lambda \quad , \quad s-i = \ell n\, \left(\frac{1-\alpha}{\beta}\right) / \ell n\, \lambda \tag{100}$$

Let T_j denote the mean number of trials from the start of the experiment and its end, if the automaton begins its motion from the state j.

If the hypothesis H_1 is true, then for T_j we obtain the finite difference equation

$$T_j = p\, T_{j+1} + r\, T_j + q\, T_{j-1} + 1 \qquad (101)$$

with boundary conditions $T_o = T_s = 0$. Ths solution of this equation has the form

$$T_j = \frac{1}{p-q} \left[j \left(\frac{1-\lambda^{-j}}{1-\lambda^{-s}} - 1 \right) + (s-j) \frac{1-\lambda^{-j}}{1-\lambda^{-s}} \right] \qquad (102)$$

Since, by hypothesis, the automaton begins its motion from the state i, then, taking into account eqs. (96), (100), we obtain

$$T_i^{(1)} = \left[(1-\alpha)\, \ell n \left(\frac{1-\alpha}{\beta} \right) + \alpha\, \ell n \left(\frac{\alpha}{1-\beta} \right) \right] / (p-q)\, \ell n\, \lambda \qquad (103)$$

If the hypothesis H_2 is true, then for T_j the equation takes the form

$$T_j = q\, T_{j-1} + r\, T_j + p\, T_{j+1} + 1 \qquad (104)$$

with the same boundary conditions. Hence

$$T_i^{(2)} = \left[\beta\, \ell n \left(\frac{1-\alpha}{\beta} \right) + (1-\beta)\, \ell n \left(\frac{\alpha}{1-\beta} \right) \right] / (p-q)\, \ell n\, \lambda \qquad (105)$$

2.9. SUPERVISED BAYES LEARNING.

The probabilities $\pi_1, \ldots\ldots, \pi_K$ that an observation belongs to class w_i, $i = 1, \ldots, K$, are assumed known. The conditional probability densities $f_i\,(\underline{z} \mid \underline{\Theta}_i) \triangleq f\,(\underline{z} \mid \underline{\Theta}_i, w_i)$ of an observation $\underline{z}$, assuming it to come from w_i, are assumed to have known functional forms, depending on parameter vectors $\underline{\Theta}_i$ some, or all of which are unknown. The problem is as follows. A sequence of generally vector-valued observations, $\underline{z}_1, \ldots, \underline{z}_n, \ldots$ are received, one at a time, and each is classified by a teacher as coming from one of a known number K of exclusive classes $w_1, \ldots\ldots, w_K$.

The problem is to learn the unknown parameters $\underline{\Theta}_i$, $i = 1, \ldots, K$, so that after an adequate training one can apply statistical decision algorithms ; sec. 2.7, for classification of new unclassified observations.

The Bayesian algorithm for learning about $\underline{\Theta}_i$ involves the specification of an a priori density $p_{io}(\underline{\Theta}_i)$ for $\underline{\Theta}_i$, and the subsequent recursive computation of the posterior density $p_i(\underline{\Theta}_i/\underline{z},\ldots,\underline{z}_n)$.

Consider what happens to the knowledge about $\underline{\Theta}_i$ when a sequence of feature vectors $\underline{z}_1,\ldots, \underline{z}_n$, all from the same pattern class w_i (that information is provided by the "teacher"), is observed. We assume that, conditional on $\underline{\Theta}_i$, the $\underline{z}_n$ are independent and identically distributed with probability density $f_i\ (\underline{z}_n/\underline{\Theta}_i)$. The function $p_{io}(\underline{\Theta}_i)$ changes to the a posteriori density function $p_i(\underline{\Theta}_i\ /\ \underline{z}_1,\ldots, \underline{z}_n)$ according to Bayes theorem. For example, the a posteriori density function of $\underline{\Theta}_i$, given the first observation, is

$$p_i\ (\underline{\Theta}_i/\underline{z}_1) = \frac{f_i(\underline{z}_1\ /\ \underline{\Theta}_i)\ p_{io}\ (\underline{\Theta}_i)}{p\ (\underline{z}_1\ /\ w_i)} \tag{106}$$

After $\underline{z}_1$ and $\underline{z}_2$ are observed the a posteriori density function of $\underline{\Theta}_i$ is

$$p_i\ (\underline{\Theta}_i\ /\ \underline{z}_1\ ,\ \underline{z}_2) = \frac{f_i(\underline{z}_2/\underline{\Theta}_i)\ p_i\ (\underline{\Theta}_i\ /\ \underline{z}_i)}{P(\underline{z}_2\ /\ \underline{z}_1\ ;\ w_i)} \tag{107}$$

In general

$$p_i\ (\underline{\Theta}_i/\underline{z}_1,\ldots, \underline{z}_n) = \frac{f_i(\underline{z}_n\ /\ \underline{\Theta}_i)\ p_i\ (\underline{\Theta}_i\ /\ \underline{z}_1,\ldots, \underline{z}_{n-1})}{P(\underline{z}_n/\underline{z}_1,\ \ldots,\ \underline{z}_{n-1}\ ;\ w_i)} \tag{108}$$

The required probability density function can be computed by

$$p(\underline{z}_n/\underline{z}_1,\ldots, \underline{z}_{n-1}\ ;\ w_i) = \int f_i\ (\underline{z}_n/\underline{\Theta}_i)\ p_i\ (\underline{\Theta}_i\ /\ \underline{z}_1,\ldots, \underline{z}_{n-1}\ ;\ w_i)\ d\underline{\Theta}_i \qquad n = 1, 2,\ldots \tag{109}$$

where the first term at the right hand side of (109), $f_i(\underline{z}_n/\underline{\Theta}_i)$ is known, and the second term is obtained from (108). The central idea of Bayesian estimation is to extract information from the observations $\underline{z}_1$, $\underline{z}_2$, ..., $\underline{z}_n$ for the unknown parameter $\underline{\Theta}_i$ through successive applications of the recursive Bayes formula (108).

The computation of equation (108) is generally overwhelming. Computation can however be rendered feasible by carefuly selecting a reproducing a priori density function for the unknown parameter so that the a posteriori density functions after each iteration are members of the same family of a priori density function (i.e., the form of the density function is preserved and only the parameters of the density function are changed). The learning schemes are then reduced to the successive estimation of parameter values.

2.10. NON - SUPERVISED BAYES LEARNING.

As in supervised learning, we assume that there are K classes of pattern vectors, and the conditional density for a given class w_i is known, except for a parameter $\underline{\Theta}_i$. In unsupervised learning, the classification of each sample in the training sequence $\underline{z}_1$, $\underline{z}_2, \ldots$, is unknown, and we wish to learn the values of $\underline{\Theta}_i$, i = 1, 2..,K, from observing the unclassified training sequence. Since the classifications of the samples are unknown, there is no way to learn separately the parameters of the density functions $f_i\ (\underline{z}/\underline{\Theta}_i)$ as we did in supervised learning.

It is intuitively obvious that unsupervised learning is more difficult than supervised learning. In practice, it is reasonable that the machine learns at first with supervision, and that after a sufficient number of iterations, the machine may learn the parameter values without supervision, and perhaps at the same time classify the observed patterns.

Unsupervised Bayes learning is formulated in terms of a mixture model. Let us define a parameter vector $\underline{\vec{\Theta}}$

$$\underline{\vec{\Theta}}^T = (\underline{\Theta}_1^T\ ,\ \underline{\Theta}_2^T, \ldots,\ \underline{\Theta}_K^T) \tag{110}$$

where T denotes transposition.

When a sample $\underline{z}$ is observed, we do not know to which class it belongs, but we may assume that it has a probability π_i of belonging to w_i. Then we may write

$$f(\underline{z}\ /\ \underline{\vec{\Theta}}) = \sum_{i=1}^{K} \pi_i\ f_i\ (\underline{z}) \tag{111}$$

where $f(\underline{z}/\underline{\vec{\Theta}})$ is the mixture density function for given $\underline{\vec{\Theta}}$, and the probabilities π_i are sometimes called mixing parameters. We have assumed that the mixing parameters π_i are known. If they are unknown, we may let

$$\underline{\vec{\Theta}}^T = (\underline{\Theta}_1^T, \ \ldots,\ \underline{\Theta}_K^T,\ \pi_1, \ldots,\ \pi_K) \tag{112}$$

and (111) still holds.

With the assumptions above, we can easily derive an unsupervised learning scheme similar to the supervised Bayes learning scheme.

Assuming that the learning observations are conditionally independent, and applying Bayes theorem,

$$P(\underline{\vec{\Theta}} / \underline{z}_1, \ldots, \underline{z}_n) = \frac{f(\underline{z}_n / \underline{\vec{\Theta}}) \, P(\underline{\vec{\Theta}} / \underline{z}_1, \ldots, \underline{z}_n)}{P(\underline{z}_n / \underline{z}_1, \ldots, \underline{z}_{n-1})}$$

$$= \frac{f(\underline{z}_n / \underline{\vec{\Theta}}) \, P(\underline{\vec{\Theta}} / \underline{z}_1, \ldots, \underline{z}_n)}{f(\underline{z}_n / \underline{\vec{\Theta}}) \, P(\underline{\vec{\Theta}} / \underline{z}_1, \ldots, \underline{z}_{n-1}) d \underline{\vec{\Theta}}} \qquad (113)$$

Substituting $f(\underline{z}_n / \underline{\vec{\Theta}})$ from (111) we get

$$P(\underline{\vec{\Theta}} / \underline{z}_1, \ldots, \underline{z}_n) = \frac{\sum_{i=1}^{K} \pi_i f_i(\underline{z}_i / \underline{\Theta}_i) \, P(\underline{\vec{\Theta}} / \underline{z}_1, \ldots, \underline{z}_n)}{\int \sum_{i=1}^{K} \pi_i f_i(\underline{z}_i / \underline{\Theta}_i) P(\underline{\vec{\Theta}} / \underline{z}_1, \ldots, \underline{z}_{n-1}) d \underline{\vec{\Theta}}} \qquad (114)$$

The relationship between $\underline{\vec{\Theta}}$ and $\underline{\Theta}_i$ is defined by (110) or (112). It is obvious that due to the mixture form inherent in (114) there exists no reproducing densities for unsupervised Bayes learning. This indicates clearly the complexity of unsupervised learning compared with supervised learning.

2.11. IDENTIFIABILITY OF FINITE MIXTURES.

In the mixture assumption for unsupervised learning, the densities $f_i(\underline{z} / \underline{\Theta}_i)$ usually belong to the same family of functions F. For example, consider the family of one-dimensional Gaussian densities with mean-value m and non-zero variance r ; $F = \{g(z, m, r), r > 0\}$. A Gaussian mixture may be written as

$$f(z/\underline{\vec{\Theta}}) = \sum_{i=1}^{K} f_i(z / \underline{\Theta}_i) \pi_i \quad , \quad f_i(z/\underline{\Theta}_i) \in F \qquad (115)$$

where

$$f_i(z / \underline{\Theta}_i) = g(z \, ; m \, , r_i) \, ,$$

$$\underline{\Theta}_i = (m_i, r_i)$$

$$\underline{\vec{\Theta}}^T = (\underline{\Theta}_1^T, \ldots, \underline{\Theta}_K^T, p_1, \ldots, p_K)$$

A major theoretical question is whether (115) is a unique representation of $f(z/\underline{\vec{\Theta}})$. In other words we ask whether there exist $\underline{\Theta}_i'$, π_i', and K' such that K and K' are finite and

$$f(z/\vec{\underline{\Theta}}) = \sum_{i=1}^{K} f_i\ (z/\underline{\Theta}_i)\ \pi_i = \sum_{i=1}^{K'} f_i\ (z/\underline{\Theta}_i')\ \pi_i' \qquad (116)$$

A trivial cause for the lack of uniqueness is that by permutation the individual terms in (115) may be labelled in K ! ways. This difficulty may be resolved by establishing an <u>ordering</u> $\prec$ in F and arranging the terms in (115) in such a way that $f_1(z/\ \underline{\Theta}_1) \prec f_2(z\ /\ \underline{\Theta}_2) \prec$ ..... For the family of Gaussian densities, we may define an ordering $g(z\ ;\ m_j,\ r_j) \prec g(z\ ;\ m_k,\ r_k)$ if $r_j > r_k$ or if $r_j = r_k$ and $m_j < m_k$. Note that, defined in this way, any subset of the Gaussian family has a unique ordering. Consider an arbitraty family F. We assume that an ordering has been defined and the densities $f_i\ (z\ /\ \underline{\Theta}_i)$ in a mixture are arranged in this order.

Under this assumption, the class of all finite mixture of F is said to be <u>identifiable</u> if (116) implies $\underline{\Theta}_i = \underline{\Theta}_i'$, $p_i = p_i'$; and $K = K'$.

The concept of identifiability was introduced by Teicher[7, 8]. Its importance to non-supervised learning is fairly obvious, since the problem is defined in terms of finite mixtures and identifiability simply means that a unique solution to the problem is possible.

2.12. PROBABILISTIC ITERATIVE METHODS - SUPERVISED LEARNING.

To formulate the problem of training an automatic recognition system mathematically, it is necessary to specify a class of possible decision functions for the system and a certain goal of learning. The goal of learning may be to attain after training the best approximation (in a certain sense) from the class of possible decision functions to a specific optimal rule of classification. In the sequel we consider the problem of classification between two classes, i.e. K = 2.

As a class of possible decision functions let us consider the functions

$$\hat{y} = \sum_{i=1}^{N} c_i\ \phi_i\ (\underline{z}) \qquad (117)$$

where c_i are unknown parameters and ϕ_i, i = 1, --, N, are a set of orthonormal functions, i.e.

$$\int_{\Omega_z} \phi_i(\underline{z})\ \phi_j(\underline{z})\ d\underline{z} = \delta_{ij}\ , \quad i,j = 1, --, N \qquad (118)$$

Here δ_{ij} denotes the kronecker deltas.

Using the vector notations

$$\underline{c}^T = (c_1, --, c_N)$$

$$\underline{\phi}^T(\underline{z}) = (\phi_1(\underline{z}), --, \phi_N(\underline{z})) \tag{119}$$

eqns. (117), (118) can be rewritten as

$$\hat{y} = \underline{c}^T \underline{\phi}(\underline{z}) \tag{120}$$

and

$$\int_{\Omega_z} \underline{\phi}(\underline{z}) \underline{\phi}^T(\underline{z}) \, d\underline{z} = I \tag{121}$$

In two-classes (w_1 and w_2) pattern recognition problem, the output $\hat{y}$ takes on either the value + 1 or -1, such that $\hat{y} = +1 \leftrightarrow \underline{z} \in \Gamma_1$, $\hat{y} = -1 \leftrightarrow \underline{z} \in \Gamma_2$. This means that $\hat{y} = +1$ corresponds to classifying $\underline{z}$ in w_1 and $y = -1$ to $\underline{z}$ in w_2.

Let us now consider the learning of different decision rules.

2.12.1. Learning Bayes rule.

It follows from eqn. (51) that the optimal discriminant function is given by :

$$g(\underline{z}) = (\nu_{12} - \nu_{11}) \pi_1 F_1(\underline{z}) + (\nu_{22} - \nu_{21}) \pi_2 F_2(\underline{z}) \tag{122}$$

such that :

$$\Gamma_1 = \{\underline{z} : g(\underline{z}) > 0\} , \Gamma_2 = \{\underline{z} : g(\underline{z}) < 0\} \tag{123}$$

The goal of learning can thus be stated as to minimize some convex function of the error between the optimal discriminant function (122) and its approximation (120). Let us consider the quadratic error function,

$$J(\underline{c}) = \int_{\Omega_z} (g(\underline{z}) - \underline{c}^T \underline{\phi}(\underline{z}))^2 \, d\underline{z} \tag{124}$$

The condition of the minimum of (124) has the form

$$\nabla J(\underline{c}) = -2 \int_{\Omega_z} (g(\underline{z}) - \underline{c}^T \underline{\phi}(\underline{z})) \underline{\phi}(\underline{z}) \, d\underline{z} = \underline{0} \tag{125}$$

Taking the orthonormality condition (121) into consideration eqn. (125) can be re-written thus

$$\underline{c} - \int_{\Omega_z} g(\underline{z})\, \underline{\phi}(\underline{z})\, d\underline{z} = \underline{0} \tag{126}$$

Substituting by $g(\underline{z})$ from (122) we get the regression equation

$$E_z \{\underline{\Phi}(\underline{z})\} = \underline{0} \tag{127}$$

where

$$\underline{\Phi}(\underline{z}) = \begin{cases} \underline{c} - (\nu_{12} - \nu_{11})\, \underline{\phi}(\underline{z}) & , \ \underline{z} \in \Gamma_1 \\ \underline{c} - (\nu_{22} - \nu_{21})\, \underline{\phi}(\underline{z}) & , \ \underline{z} \in \Gamma_2 \end{cases} \tag{128}$$

Applying probabilistic iterative methods the following algorithm is obtained for the current estimate of the solution of the regression eqn. (127),

$$\begin{aligned} \underline{c}(n) &= \underline{c}(n-1) - n^{-1}\left[\underline{c}(n-1) - (\nu_{12} - \nu_{11})\, \underline{\phi}(\underline{z}(n))\right] \text{ if } y(n)=+1 \\ \underline{c}(n) &= \underline{c}(n-1) - n^{-1}\left[\underline{c}(n-1) - (\nu_{22} - \nu_{21})\, \underline{\phi}(\underline{z}(n))\right] \text{ if } y(n)=-1 \end{aligned} \tag{129}$$

The block diagram of learning system that realizes this algorithm is shown in Fig. 5.

2.12.2. <u>Learning Seigert-Kotelnikov (also max. a posteriori) Rule</u>.

The learning algorithm can be obtained as a particular case of (129) when

$$\nu_{11} = \nu_{22} = 0 \ ; \ \nu_{12} = \nu_{21} = 1 \tag{130}$$

This yields the learning algorithm

$$\begin{aligned} \underline{c}(n) &= \underline{c}(n-1) - n^{-1} [\underline{c}(n-1) - \underline{\phi}(\underline{z}\ (n))] , \quad \text{if } y(n) = +1 \\ \underline{c}(n) &= \underline{c}(n-1) - n^{-1} [\underline{c}(n-1) + \underline{\phi}(\underline{z}\ (n))] , \quad \text{if } y(n) = -1 \end{aligned} \tag{131}$$

Or what amounts to the same,

$$\underline{c}(n) = \underline{c}(n-1) - n^{-1} [\underline{c}(n-1) - y(n)\, \underline{\phi}(\underline{z}(n))] \tag{132}$$

The block, diagram of learning system that realizes this algorithm is shown in Fig.6.

2.12.3. Learning Mixed Decision Rule.

In this case

$$\nu_{11} = \nu_{22} = 0 \;;\; \nu_{12} = 1 \;,\; \nu_{21} = \mu \tag{133}$$

This yields the learning algorithm,

$$\begin{aligned} \underline{c}(n) &= \underline{c}(n-1) - n^{-1}\,[\underline{c}(n-1) - \underline{\phi}(\underline{z}(n))] \;, && \text{if } y(n) = +1 \\ \underline{c}(n) &= \underline{c}(n-1) - n^{-1}\,[\underline{c}(n-1) + \mu\underline{\phi}(\underline{z}(n))] \;, && \text{if } y(n) = -1 \end{aligned} \tag{134}$$

The corresponding block diagram is shown in Fig.7.

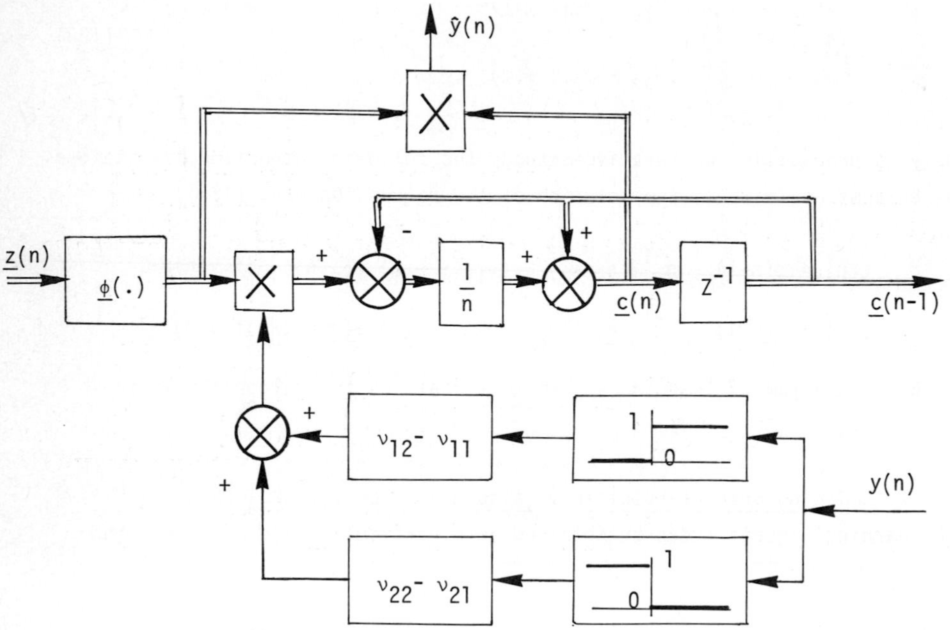

Fig.5. Supervised learning of Bayes Rule

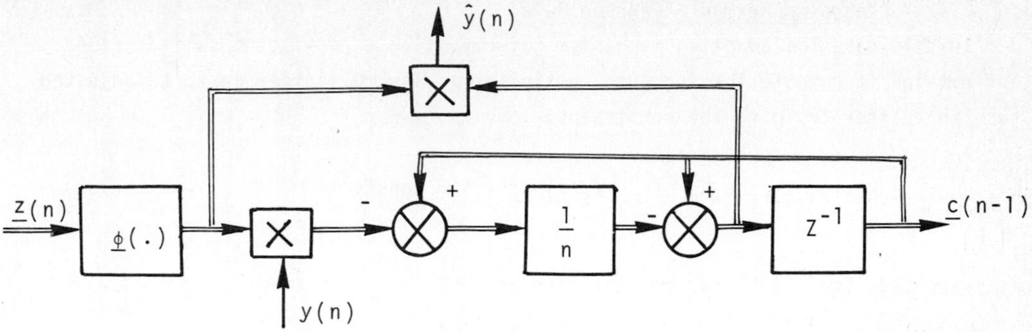

Fig.6 - Supervised Learning of Siegert - Kotelnikov rule

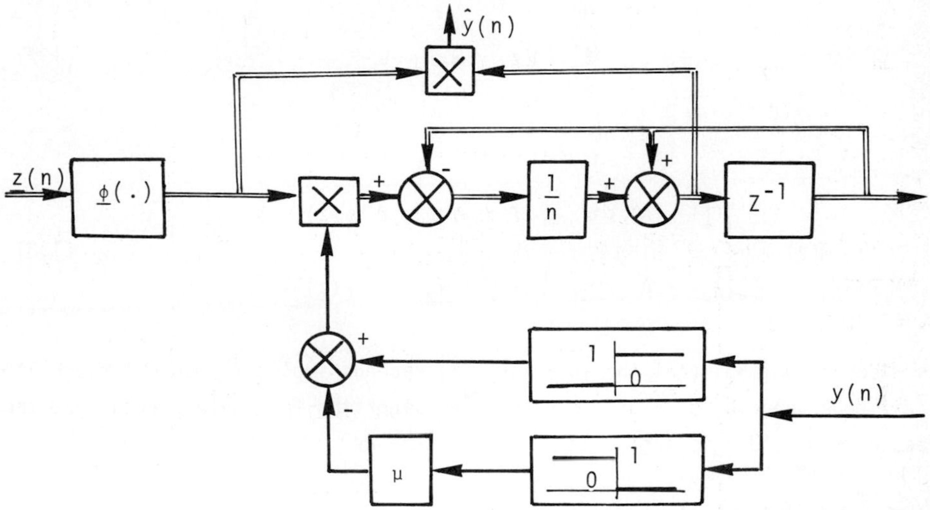

Fig.7 - Supervised Learning of mixed decision rule

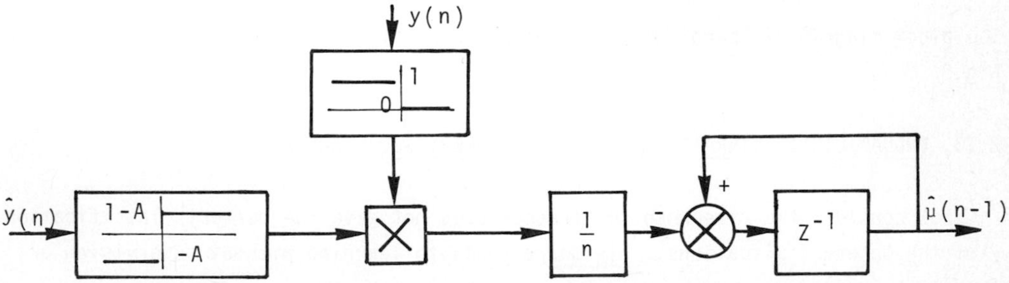

Fig.8 - Supervised Learning of μ for Neyman - Pearson rule

2.12.4. Learning Neyman - Pearson Rule.

In this case the adaptive mechanism of computing $\underline{c}$ (n) is the same as (134) but remains to compute the lagrange multiplier μ. The multiplier is to be adjusted such that after training the constraint,

$$\int_{\Gamma_1} f_2(\underline{z})\, d\underline{z} = A = \text{const.} \tag{135}$$

is respected. Eqn. (135) may be rewritten as

$$\int_{\Gamma_1} (1 - A)\, f_2(\underline{z})\, d\underline{z} - \int_{\Gamma 2} A\, f_2(\underline{z})\, d\underline{z} = 0 \tag{136}$$

which amounts to the regression equation

$$E_{\underline{z}}\ \{\Theta(\underline{z}) \;/\; w_2\} = 0 \tag{137}$$

where

$$\Theta(\underline{z}) = \begin{cases} (1 - A) \;/\; \pi_2 \;, \text{if } \underline{z} \sim w_2 \;, \; \underline{c}^T \underline{\phi}(\underline{z}) > 0 \\ - A \;/\; \pi_2 \;, \text{if } \underline{z} \sim w_2 \;, \; \underline{c}^T \underline{\phi}(\underline{z}) < 0 \end{cases} \tag{138}$$

It is clear from (138) that the function $\Theta(\underline{z})$ depends implicitly on μ through its explicit dependence on $\underline{c}$. A discrete learning algorithm to update μ such that the regression eqn. (137) is satisfied may be written as

$$\begin{aligned} \mu(n) &= \mu(n-1) - n^{-1}(1-A) \;/\; \pi_2 \;, \quad \text{if } y(n) = -1 \;, \; \underline{c}^T \underline{\phi}(\underline{z}) > 0 \\ \mu(n) &= \mu(n-1) + n^{-1}\, A \;/\; \pi_2 \;, \quad \text{if } y(n) = -1 \;, \; \underline{c}^T \underline{\phi}(\underline{z}) < 0 \end{aligned} \tag{139}$$

The block diagram of learning system that realizes the algorithms (139) is shown in Fig.8.

2.13. PROBABILISTIC ITERATIVE METHODS - UNSUPERVISED LEARNING.

Now consider the case when the teacher does not give the correct classification y of the observed situations. This corresponds to learning without supervision or to self-learning.

Let the goal of the self-learning system is to learn the Siegert-Kotelnikov maximum a posteriori probability decision rule, where the discriminant function is

$$g(\underline{z}) = \pi_2 f_2(\underline{z}) - \pi_1 f_1(\underline{z}) \qquad (140)$$

Let us assume now that the products of a priori probabilities and conditional density functions $\pi_1 f_1(\underline{z})$ and $\pi_2 f_2(\underline{z})$ can be approximated by a finite series

$$\pi_2 f_2(\underline{z}) \simeq \underline{a}^T \underline{\Phi}(\underline{z}) \quad , \quad \pi_1 f_1(\underline{z}) \simeq \underline{b}^T \underline{\Psi}(\underline{z}) \qquad (141)$$

Here, $\underline{a}^T = (a_1, --, a_{N_1})$; $\underline{b}^T = (b_1, --, b_{N_2})$ are unknown vectors, and

$$\underline{\Phi}^T(\underline{z}) = (\Phi_1(\underline{z}), --, \Phi_{N_1}(\underline{z})) \ , \ \underline{\Psi}^T(\underline{z}) = (\Psi_1(\underline{z}), --, \Psi_{N_2}(\underline{z}))$$

are known vectors functions. For simplicity, their component functions are assumed to form an orthonormal system.

The decision rule (140) can then be written in the form

$$\hat{g}(\underline{z}, \underline{a}, \underline{b}) = \underline{a}^T \underline{\Phi}(\underline{z}) - \underline{b}^T \underline{\Psi}(\underline{z}) \qquad (142)$$

and the decision rule is determined by finding the vectors $\underline{a}$ and $\underline{b}$. But these vectors can be found in the following manner. Noticing that due to (141) the probability density function

$$f(\underline{z}) = \pi_1 f_1(\underline{z}) + \pi_2 f_2(\underline{z}) \qquad (143)$$

is approximately equal to

$$f(\underline{z}) \simeq \underline{a}^T \underline{\Phi}(\underline{z}) + \underline{b}^T \underline{\Psi}(\underline{z}) \qquad (144)$$

it is simple to understand that the problem of determining the vectors $\underline{a}$ and $\underline{b}$ is reduced to the restoration (estimation) of the mixture probability density function.

Let us introduce the functional

$$J(\underline{a}, \underline{b}) = \int_{\Omega_z} (f(\underline{z}) - \underline{a}^T \underline{\Phi}(\underline{z}) - \underline{b}^T \underline{\Psi}(\underline{z}))^2 \, d\underline{z} \qquad (145)$$

By differentiating this functional with respect to $\underline{a}$ and $\underline{b}$, and considering the orthonormality of the component functions $\underline{\Phi}(\underline{z})$ and $\underline{\Psi}(\underline{z})$, we find the conditions of the minimum in the form

$$\nabla_{\underline{a}} J(\underline{a}, \underline{b}) = E\{\underline{\Phi}(\underline{z})\} - \underline{a} - G\underline{b} = \underline{0}$$
$$\nabla_{\underline{b}} J(\underline{a}, \underline{b}) = E\{\underline{\Psi}(\underline{z})\} - G^T\underline{a} - \underline{b} = \underline{0} \tag{146}$$

where the matrix

$$G = \int_{\Omega_z} \underline{\Phi}(\underline{z}) \Psi^T(\underline{z}) \, d\underline{z} \tag{147}$$

By solving eqs. (146) with respect to $\underline{a}$ and $\underline{b}$, we obtain

$$\underline{a} = E \{ U (\underline{\Phi}(\underline{z}) - G \underline{\Psi}(\underline{z})) \} , \tag{148}$$

$$\underline{b} = E \{ U (\underline{\Psi}(\underline{z}) - G^T\underline{\Phi}(\underline{z})) \} , \tag{149}$$

where $U = (I - \underline{G} \, \underline{G}^T)^{-1}$.

The simplest optimal stochastic approximation algorithms for solving the regression equations (148) and (149) are,

$$\underline{a}(n) = \underline{a}(n-1) - n^{-1} (\underline{a} (n-1) - U(\underline{\Phi}(\underline{z}(n)) - G\underline{\Psi}(\underline{z}(n))))$$
$$\underline{b}(n) = \underline{b}(n-1) - n^{-1} (\underline{b} (n-1) - U(\underline{\Psi} (\underline{z}(n)) - G^T \underline{\Phi}(\underline{z}(n)))) \tag{150}$$

The learned decision rule will have the form

$$\hat{g}(\underline{z}(n) \, , \underline{a} (n-1) \, , \underline{b} (n-1)) = \underline{a}^T(n-1) \, \underline{\Phi} (\underline{z}(n)) - \underline{b}^T(n-1) \, \underline{\Psi}(\underline{z}(n)) \tag{151}$$

The block diagram of the self-learning system that uses these algorithms is shown in Fig.9.

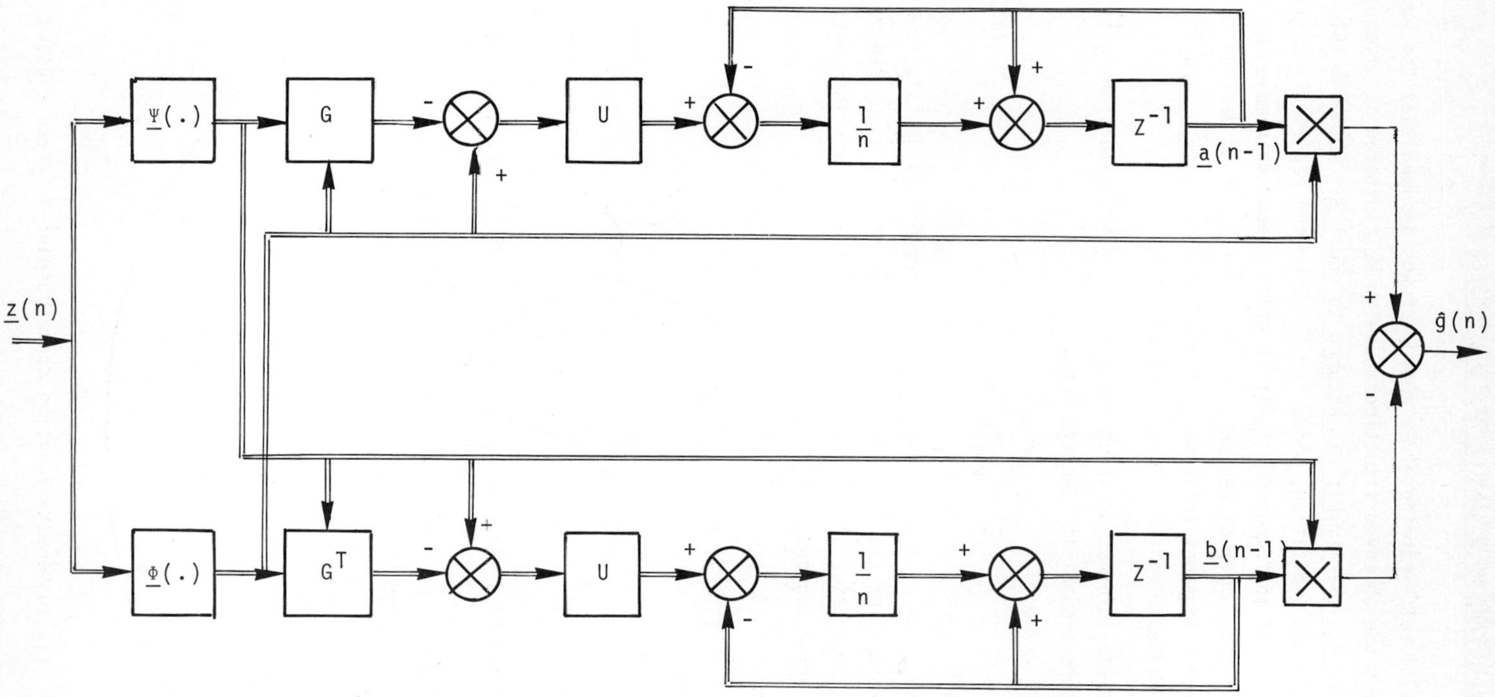

Fig.9. Non-supervised learning - Parametric expansion.

2.14. SELF LEARNING WITH UNKNOWN NUMBER OF PATTERN CLASSES.

In the algorithms of self-learning given above, it was assumed that the number of regions K into which the observed situations have to be clustered is given in advance (for simplicity and clarity, it was assumed to equal 2). Although this does not look like a significant limitation, since for $K > 2$ we can repeatedly use the binary case (frequently called "dichotomy"), it is still needed to remove the necessity of specifying a fixed number of regions. In other words, it is desired not only to relate observed situations to proper regions but also to determine the correct number of these regions.

Sufficiently complete information about the regions of the situations $\underline{z}$ is contained in the mixture probability density function

$$f(\underline{z}) = \sum_{k=1}^{K} \pi_k \, f_k(\underline{z}) \tag{152}$$

We can assume that the peaks of the estimated mixture probability density function correspond to the "centers" of the regions, and the lines passing along the valleys of its relief are the boundaries of the regions ; the number of existing peaks in $f(\underline{z})$ defines the number of regions, see Fig.10.

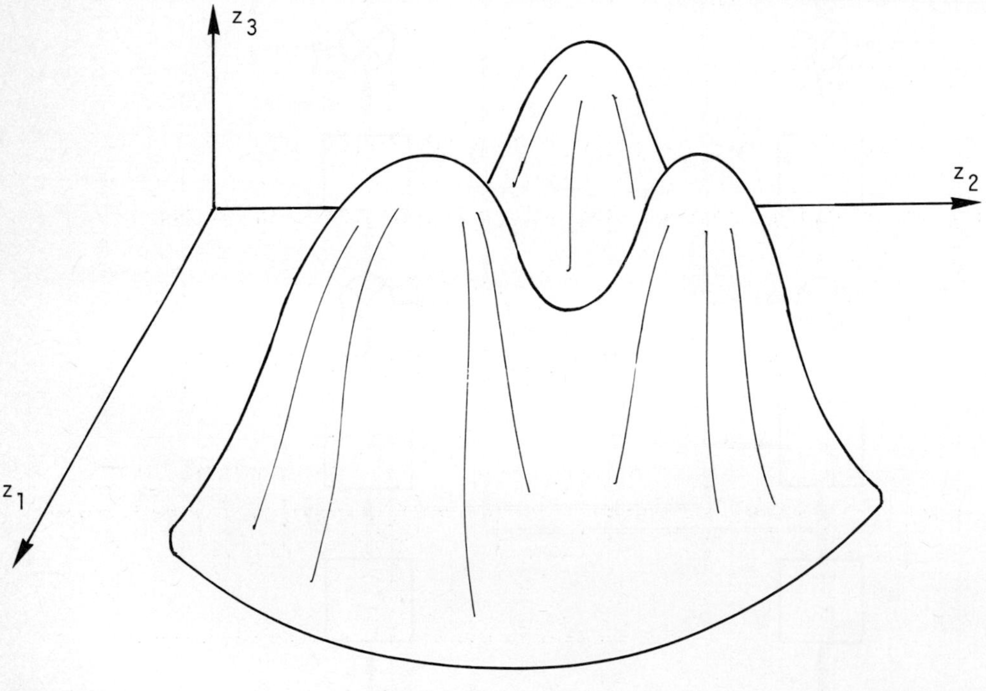

Fig.10 Density function corresponding to three classes.

In order to restore (estimate) the mixture probability density function $f(\underline{z})$ we shall approximate it by

$$f(\underline{z}) = \underline{a}^T \underline{\Phi}(\underline{z}) \tag{153}$$

where $\underline{\Phi}(\underline{z})$ is a vector function with orthonormal components.

We now form the functional

$$J(\underline{a}) = \int_{\Omega_z} (f(\underline{z}) - \underline{a}^T \underline{\Phi}(z))^2 \, d\underline{z} \tag{154}$$

for which the necessary condition for optimality leads to the regression equation

$$\underline{a} = E\{\underline{\Phi}(\underline{z})\} \tag{155}$$

A probabilistic iterative algorithm for solving (155) may be written as

$$\underline{a}(n) = \underline{a}(n-1) - n^{-1} (\underline{a}(n-1) - \underline{\Phi}(\underline{z}(n))) \tag{156}$$

The algorithm (156) is an optimal one [9].

According to (153),

$$\hat{f}_n(\underline{z}(n)) = \underline{a}^T(n)\underline{\Phi}(\underline{z}(n)) \tag{157}$$

The system realizing algorithms (156) and (157) is presented in Fig.11.

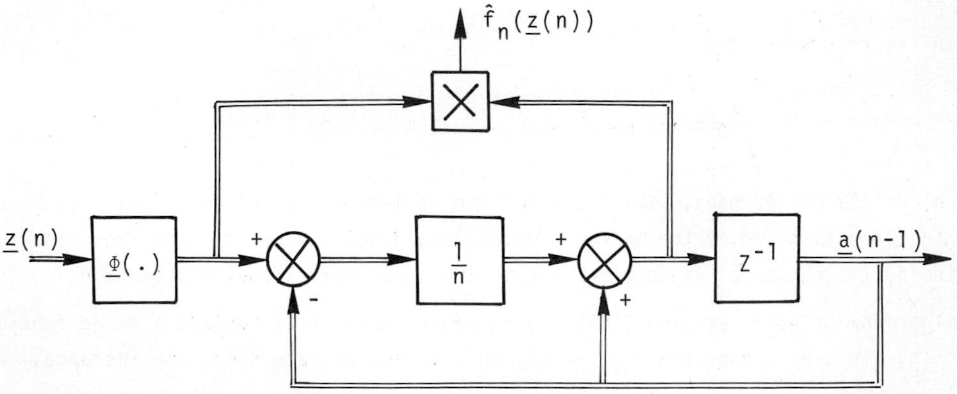

Fig.11 Learning the mixed density function.

Therefore, we can form an estimate of the mixture probability density function.

A slightly different approach to restoration (estimation) of $f(\underline{z})$ is also possible. It may be practical to define the mixture probability density function using the estimator proposed by Rosenblatt[20],

$$f_n(\underline{z}) = \frac{1}{2nh} \sum_{m=1}^{n} I_{(-1,\ 1)} \left(\frac{\underline{z} - \underline{z}(m)}{h}\right) \tag{158}$$

Where $I_A(\ .\ )$ is the characteristic function

$$\begin{aligned} I_A(\underline{x}) &= 1 \qquad \text{if } \underline{x} \in A \\ &= 0 \qquad \text{otherwise} \end{aligned} \tag{159}$$

Rosenblatt[20] demonstrated the convergence (in mean square) of $f_n(\underline{z})$ towards $f(\underline{z})$ on the condition that h is a function of n such that $h_n \to 0$ as $n \to \infty$ with h_n converging to zero slower than $\frac{1}{n}$.

We note that choosing the I function in (158) yields a contribution of the "needle-type" following each observation.

It may be preferred to replace the I - function,eqn. (158),by a certain bell-shaped function $\eta(\underline{z}, \underline{z}(m))$ (Fig.12). That gives the largest weight to the observed situation $\underline{z}(m)$, and for the other situation, the weights are different from zero. This yields smooth estimates of the density function. Then instead of (158) we obtain

$$f_n(\underline{z}) = n^{-1} \sum_{m=1}^{n} \eta(\underline{z}, \underline{z}(m)) \tag{160}$$

or, in the recursive form,

$$f_n(\underline{z}) = f_{n-1}(\underline{z}) - n^{-1}(f_{n-1}(\underline{z}) - \eta(\underline{z}, \underline{z}\ (n))) \tag{161}$$

This algorithm of learning, like the algorithm of learning (156) and (157), can be used in the estimation of the mixture probability density function, and thus also in finding the number of regions or classes and their corresponding situations.

The algorithm of self-learning (161) can be generalized if we replace a fixed function $\eta(\underline{z}, \underline{z}(n))$ by a function $\eta_n(\underline{z}, \underline{z}(n))$ that varies at each step, for instance,

$$\eta_n\ (\underline{z}, \underline{z}(n)) = (h(n))^{-1}\ \eta\left(\frac{\underline{z} - \underline{z}(n)}{h(n)}\right) \tag{162}$$

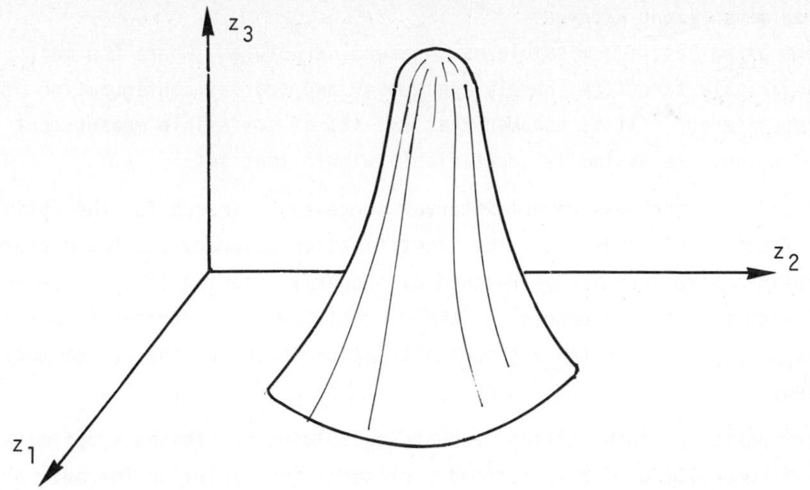

Fig. 12 Bell-Shaped function.

where h(n) is a certain decreasing sequence of positive numbers. Eqn. (162) has the meaning that the distributions get "sharpened" around the centers as n increases. So their effets become secondary ; they merely contribute "needle" changes (corresponding to δ - function) as $n \to \infty$.

It should be noticed that the algorithms of learning (156) and (157) are the special cases of the lagorithm of learning (161). Actually, by setting

$$\eta(\underline{z}, \underline{z}(n)) = \underline{\Phi}^T(\underline{z})\underline{\Phi}(\underline{z}(n)) \qquad (163)$$

in (161), and by introducing $f_k(\underline{z})$ from (157), we obtain the algorithm of learning (156) after a division by $\underline{\Phi}(\underline{z})$.

We have described above the way toward the restoration (estimation) of the mixture probability density functions. For multidimensional vectors of the situation $\underline{z}$, this restoration is very difficult when smoothness has to be maintained. It is even more difficult to extract the desired regions.

2.15. APPLICATION - MEASUREMENT STRATEGY FOR SYSTEMS IDENTIFICATION.

In some pratical situations, it is required to identify a dynamic system under the following limitations :

i - fixed measurement interval,

ii - constrained set of admissible measurement structures, where the measurement system has a variable structure, namely the number and spatial configuration of the sensors can be altered[14]. It is assumed that the set of admissible measurement structures is finite, and the system is identifiable within that set.

The limitation on the measurement interval suggests to search for the optimal measurement structure by trading off the identification accuracy and the measurement cost. That optimal structure may be reached by properly altering the measurement structure at each time step depending on the current level of uncertainties (or covariance errors). The algorithm for such alteration is called the system measurement strategy.

The determination of such strategy amounts to solving the feedback optimization problem. It is generally difficult to find a closed - form solution for such problem. On the other hand, a numerical solution is hindered by the so-called "curse of optimality". Even for linear dynamic systems, the open-loop numerical solution-corresponding to a particular initial uncertainty - is so a formidable task that its on-line implementation is practically impossible[15].

In order to deal with such difficulty, we propose here the application of pattern recognition techniques for the determination of these optimal measurement strategies. The form of the present solution admits on-line application, and may be equally applied to non-linear systems.

2.15.1. Problem formulation.

Let the system dynamics be described by the following discrete-time recurrence equation :

$$\underline{x}(n+1) = \underline{f}_n(\underline{x}(n)) + \underline{w}(n) \tag{164}$$

where $\underline{f}_n(.)$ is a known non linear vector function ; $\underline{x}(n)$ and $\underline{w}(n)$ denote, respectively, the state and disturbance vectors, at the time step n = 0, 1, ..., N-1. The vectors $\underline{f}_n$, $\underline{x}$ and $\underline{w}$ are all p-dimensional.

The disturbance sequence $\{\underline{w}(n)\}$ is assumed to be constituted of independent stationary random variables of zero mean and known covariance,

$$E\{\underline{w}(n)\ \underline{w}^T(n)\} = V_w(n) \tag{165}$$

The system initial state is known only within certain a priori statistics assumed to be Gaussian with the following mean and covariance

$$E\{\underline{x}(0)\} = \underline{\hat{x}}(0), \quad E\{(\underline{x}(0) - \underline{\hat{x}}(0))(\underline{x}(o) - \underline{\hat{x}}(0)^T\} = V_{\tilde{x}}(0) \tag{166}$$

It is supposed that the system measurements at time n are represented by the r-dimensional ($r \leqslant p$) vector given by

$$\underline{z}(n) = \underline{g}_n(\underline{x}(n), \underline{c}(n)) + \underline{v}(n) \tag{167}$$

where $\underline{v}$ is the measurement noise. The sequence $\{\underline{v}(n)\}$ is assumed to be independent of $\{\underline{w}(n)\}$ and also constituted of independent stationary random variables of zero mean and known convariance

$$E\ \{\underline{v}(n)\ \underline{v}^T(n)\} = V_v(n) \tag{168}$$

The vector $\underline{c}(n)$ specifies the measurement structure, which characterizes the relationship between the system state-parameter vector and the measurement vector at time step n. Such measurement structure has to be a member of the set of admissible measurement structures,

$$C = \{\underline{c}^1, \underline{c}^2, \ldots, \underline{c}^M\} \tag{169}$$

i.e., at time step n, $\underline{c}(n)$ can take any value $\underline{c}^i$, $i = 1, \ldots, M$.

The cost of identification errors is a function of the covariance matrix $V_{\tilde{x}}(N)$ at the terminal time step N. Let us denote that function by $\phi\{V_{\tilde{x}}(N)\}$. The measurement cost depends solely on the measurement structures $\underline{c}(n)$, $n = 0, 1, \ldots, N-1$, and can be expressed by the summation

$$\sum_{n=0}^{N-1} \psi_n[\underline{c}(n)] \tag{170}$$

Hence, the problem is to specify the measurement structure at each time step, i.e. to determine the optimal strategy $\underline{c}^*(n)$, $n = 0, 1, \ldots, N-1$ such as to minimize the overall cost

$$Q(0, N) = \phi\{V_{\tilde{x}}\ (N)\} + \lambda \sum_{n=0}^{N-1} \psi_n[\underline{c}(n)] \tag{171}$$

where $\lambda \geqslant 0$ is a weighting factor compromizing the identification accuracy (first term) and the measurement cost (second term).

2.15.2. Extended Kalman filter[16].

Let us suppose the optimal measurement structure $\underline{c}^*(m)$ has been determined for all m, m = 0, 1, ..., n-1. Now introduce the matrices

$$F(n) = \frac{\partial \underline{f}_n(\hat{\underline{x}}(n))}{\partial \hat{\underline{x}}(n)} \quad , \quad G(n+1) = \frac{\partial \underline{g}_{n+1}(\hat{\underline{x}}((n+1)/n), \underline{c}(n+1))}{\partial \hat{\underline{x}}((n+1)/n)} \tag{172}$$

the a priori variance matrix

$$\underline{V}_{\tilde{x}}((n+1)/n) = F(n)\, V_{\tilde{x}}(n)\, F^T(n) + V_w(n) \tag{173}$$

and the gain matrix

$$K(n+1) = V_{\tilde{x}}((n+1)/n)\, G^T(n+1)\, [G(n+1)\, V_{\tilde{x}}((n+1)/N)\, G^T(n+1) + V_v(n+1)]^{-1} \tag{174}$$

$\hat{\underline{x}}((n+1)/n)$ in (172) denotes the one-stage prediction state determined by

$$\hat{\underline{x}}((n+1)/n) = \underline{f}_n(\hat{\underline{x}}(n)) \tag{175}$$

The filter equations can then be written as

$$\hat{\underline{x}}(n) = \hat{\underline{x}}((n+1)/n) + K(n+1)\, [\underline{z}(n+1) - \underline{g}_n(\hat{\underline{x}}((n+1)/n), \underline{c}(n+1))] \tag{176}$$

The covariance matrix for that estimate is given by the algorithm

$$V_{\tilde{x}}(n+1) = [I-K(n+1)\, G(n+1)]\, V_{\tilde{x}}((n+1)/n)\, [I-K(n+1)\, G(n+1)]^T + $$
$$+ K(n+1)\, V_v(n+1)\, K^T(n+1) \tag{177}$$

Due to the symmetry of the matrix $V_{\tilde{x}}(n+1)$ it can be represented by a vector $\alpha(n+1)$

$$\underline{\alpha}(n+1)^T = (\alpha_1(n+1), \alpha_2(n+1), \ldots, \alpha_{p'}(n+1)) \; , \; p' = \frac{p(p+1)}{2} \tag{178}$$

$$\alpha_i(n+1) = (V_{\tilde{x}}(n+1))_{j,k} \tag{179}$$

where

$$\begin{array}{llr} j = k = i & \text{for } i = 1, 2, \ldots, p & \\ j = k - 1 = i - p & \text{"} \quad i = p + 1, \ldots, 2p-1 & (180) \\ j = k - 2 = i - (2p - 1) & \text{"} \quad i = 2p, \ldots, 3p-3 & \\ \vdots & \vdots & \\ j = k - (p-1) = i - \frac{p(p+1)}{2} + 1 & \text{"} \quad i = \frac{p(p+1)}{2} & \end{array}$$

Thus the up-right diagonals of the $V_{\tilde{x}}$ matrix are arranged componentwise for the $\underline{\alpha}$ vector. Henceforth, the vector α will be referred to as the variance vector.

Let us note that the variance vector $\underline{\alpha}(n)$ at time n is a function of $\underline{\alpha}(0)$ and the measurement structures $\underline{c}(m)$ for all m = 0, 1, ..., n-1 through the sequential representation

$$\underline{\alpha}(n) = \underline{h}_{n-1}\{\underline{\alpha}(n-1), \underline{c}(n-1)\}, \qquad n = 1, 2, \ldots N$$
$$\underline{\alpha}(0) \text{ given} \tag{181}$$

Here $\underline{h}_{n-1}$ denotes the algorithm given by (172), (173), (174) and (177).

In terms of the variance vector $\underline{\alpha}(n)$ the overall cost (171) can be written as

$$Q(0,N) = \phi(\underline{\alpha}(n)) + \lambda \sum_{n=0}^{N-1} \psi_n[\underline{c}(n)], \qquad \lambda \geqslant 0 \tag{182}$$

2.15.3. Dynamic programming solution.

Consider the time step N-1. It follows from (181) that

$$\phi[\underline{\alpha}(N)] = \phi[\underline{h}_{N-1}\{\underline{\alpha}(N-1), \underline{c}(N-1)\}] \tag{183}$$

Assume that $\underline{c}^*(0), \ldots, \underline{c}^*(N-2)$ are determined, then the optimization of (182) will be reduced to minimizing

$$Q(N-1,N) = \phi[\underline{h}_{N-1}(\underline{\alpha}(N-1), \underline{c}(N-1))] + \lambda\, \psi_{N-1}[\underline{c}(N-1)] \tag{184}$$

over the set of all admissible measurement structures C. Fixing numerical values for $\underline{\alpha}(N-1)$ will then make (184) a function of $\underline{c}(N-1)$, which can be minimized over the set C to yield a value $\underline{c}^*(N-1)$ corresponding to the numerical value assigned to $\underline{\alpha}(N-1)$. That procedure is repeated for large different possible realizations of $\underline{\alpha}(N-1)$ which can be produced by suitable random generation. The numerical results are then tabulated according to patterns or classes of elements. In that respect one may define M patterns : $A^1, \ldots, A^i, \ldots, A^M$ the ith of which can be represented as

$$A^i = \{\underline{\alpha}^j : c^*(\underline{\alpha}^j) = \underline{c}^i\} \quad , \quad i = 1, \ldots, M \tag{185}$$

Here $\underline{c}^*(\underline{\alpha}^j)$ denotes the optimal structure corresponding to a value $\underline{\alpha}^j$. $\underline{c}^i$ is an element of the set C. Notice that there are M classes of $\underline{\alpha}$ (at any time step n) which equals the cardinal number of the set C.

Let us denote by $\underline{\rho}_n$ the decision vector function needed to recognize the appropriate pattern of a sample $\underline{\alpha}(n)$ at time n. By means of that decision function it is possible to assign the optimal structure $\underline{c}^*$ corresponding to the variance vector $\underline{\alpha}$ at time n. Let us represent that decision procedure formally by the equation

$$\underline{c}^*(n) = \underline{\zeta}_n(\underline{\rho}_n, \underline{\alpha}(n)) \quad , \quad n = N - 1, \ldots, 1, 0 \tag{186}$$

The algorithm corresponding to equation (186) will be discussed in the following section.

Notice that the time step is considered to be n rather than N-1 in (186) in order to save writing when the procedure is again mentioned in the following arguments.

Substituting by (186) (for n = N-1) into (184) we get the optimal one-stage cost

$$Q(N-1,N) = \phi\,[\underline{h}_{N-1}(\underline{\alpha}(N-1), \underline{\zeta}_{N-1}(\underline{\rho}_{N-1}, \underline{\alpha}(N-1)))] +$$

$$+\lambda\ \psi_{N-1}[\underline{\zeta}_{N-1}(\underline{\rho}_{N-1}, \underline{\alpha}(N-1))] = \gamma_{N-1}[\underline{\alpha}(N-1)] \tag{187}$$

It is to be emphasized that γ_{N-1} denotes a numerical procedure rather than a function. An analytical expression of γ_{N-1} is not needed.

Let us now consider the two stage decision process (N-2,N) for which the cost can be written as

$$Q(N-2,N) = \gamma_{N-1}[\underline{\alpha}(N-1)] + \lambda\ \psi_{N-2}(\underline{c}(N-2))$$

$$= \gamma_{N-1}[\underline{h}_{N-1}(\underline{\alpha}(N-2), \underline{c}(N-2))] + \lambda\ \psi_{N-2}(\underline{c}(N-2)) \tag{188}$$

By analogy with (184) it is obvious that optimization of Q(N-2,N) can proceed in principle as the same as before. The numerical results (relating $\underline{\alpha}(N-2)$ and $\underline{c}^*(N-2)$ thus obtained can then be classified in M patterns (185). The appropriate decision procedure (186) is then sought and so on.

2.15.4. Pattern recognition.

Let us consider the classification of the M patterns (185) by successive dichotomy[4], i.e. by successively splitting the patterns (or classes) into groups.

If we begin, for example, with four classes ; we could solve the following two problems in the order given :

(1) Find a decision boundary separating classes 1 and 2 from classes 3 and 4.
(2) Using samples of classes 1 and 2 alone, find a decision boundary separating class 1 from 2. Using samples of classes 3 and 4, find a decision boundary separating class 3 and 4.

Notice that the decision procedure need to be carried out a maximum of M-1 times for an M-class problem. This can easily be shown Fig.13 where the decision procedure is depicted as tree structure, each of the nodes corresponds to a decision function. The parallel structure (Fig.13.a) has the advantage of quick decisions since the maximum number of decision functions in a decision procedure is less than that of the series structure (Fig.13.b).

The determination of a component of the $\underline{\rho}$ vector function separating two classes can be established by the probabilistic iterative approach[9].

Let us designate the decision function by

$$\hat{y} = \rho_i(\underline{\alpha},\underline{S}) = \underline{S}^T \underline{\tau}(\underline{\alpha}) \tag{189}$$

where $\rho_i(\underline{\alpha},\underline{S})$ is a function that is known up to the parameter vector $\underline{S}^T = (s_1,\ldots,s_\nu)$. The signs of the decision functions define the regions §

$$\chi_1 = \{\underline{\alpha} : \rho_i(\underline{\alpha},\underline{S}) < 0\} \quad , \quad \chi_2 = \{\underline{\alpha} : \rho_i(\underline{\alpha},\underline{S}) > 0\} \tag{190}$$

On the other hand, a teacher gives us the correct classification of each observed sample $\underline{\alpha}$:

$$y = \begin{cases} -1 & \text{if } \underline{\alpha} \text{ is classified into } \chi_1^\circ \\ +1 & \text{if } \underline{\alpha} \text{ is classified into } \chi_2^\circ \end{cases} \tag{191}$$

The superscript ° denotes the correct class. Obviously the decision will be correct

§ Notice that according to Fig.13.a or b the classes χ_1 or χ_2 may consist of the summation of several patterns $A_i,\ldots, A_j$ according to the decision tree and the decision node.

if

$$y.\rho_i(\underline{\alpha},\underline{S}) > 0 \tag{192}$$

and incorrect if the opposite is true, i.e.

$$y.\ \rho_i(\underline{\alpha},\ \underline{S}) < 0 \tag{193}$$

As a penalty function for misclassification, we select a certain convex function of the difference between y and $\hat{y}$, that is

$$D\ (\ y - \rho_i(\underline{\alpha},\underline{S})) \tag{194}$$

Then the average risk of misclassification can be written as

$$\begin{aligned} R &= \int D\ (\ y - \rho_i(\underline{\alpha},\underline{S}))\ \ p(\underline{\alpha})\ d\underline{\alpha} \\ &= E_{\underline{\alpha}}\ \{\ D\ (\ y - \rho_i(\underline{\alpha},\underline{S}))\ \ \} \end{aligned} \tag{195}$$

where $p(\underline{\alpha})$ is the unknown mixture probability density

$$p(\underline{\alpha}) = \sum_{k=1}^{M} P_i\ p_i(\underline{\alpha}) \tag{196}$$

We denote by $P_1, \ldots, P_M$ the probabilities of occurence of $\underline{\alpha}$ in the classes $\chi_1, \ldots, \chi_M$; and by $p_1(\underline{\alpha}), \ldots, p_M(\underline{\alpha})$ the conditional probability densities of $\underline{\alpha}$ in classes $\chi_1, \ldots, \chi_M$.

The necessary condition for the minimum of the average risk is

$$\begin{aligned} \nabla_{\underline{S}}\ R &= \int \nabla_{\underline{S}}\ D\ (\ y - \rho_i(\underline{\alpha},\underline{S}))\ \ p(\underline{\alpha})\ d\underline{\alpha} \\ &= E_{\underline{\alpha}}\ \{\ \nabla_{\underline{S}}\ D\ (\ y - \rho_i(\underline{\alpha},\underline{S})\ )\} = \underline{0} \end{aligned} \tag{197}$$

Now on the basis of the results of [9] we can easily obtain the learning algorithms for the pattern recognition system

$$\underline{S}(k) = \underline{S}(k-1) + \Gamma(k)\ D'(y(k) - \underline{S}^T(k-1)\ \underline{\tau}\{\underline{\alpha}(k)\})\ \underline{\tau}\{\underline{\alpha}(k)\}\ ^{\S} \tag{198}$$

---

§ $D'(x) = dD(x)\ /\ dx$

where $\Gamma(k)$ is a step-length matrix to be defined.

If the function D(.) is considered to be the quadratic function

$$D\ (y - \rho_i(\underline{\alpha},\underline{S})\ ) = \frac{1}{2}\,(y - \rho_i(\underline{\alpha},\underline{S}))^2 \qquad (199)$$

then the algorithm for optimal learning [9] will be

$$\underline{S}(k) = \underline{S}(k-1) + E(k)\ [y(k) - \underline{S}^T(k-1)\ \underline{\tau}\{\underline{\alpha}(k)\}]\ \underline{\tau}\{\underline{\alpha}(k)\}$$

$$E(k) = E(k-1) - \frac{E(k-1)\ \underline{\tau}\{\underline{\alpha}(k)\}(E(k-1)\ \underline{\tau}\{\underline{\alpha}(k)\})^T}{1 + \underline{\tau}^T\{\underline{\alpha}(k)\}\ E(k-1)\ \underline{\tau}\{\underline{\alpha}(k)\}} \qquad (200)$$

$$S(0)\ \text{arbitrary, and}\ E(0) > 0.$$

2.15.5. On-line identification.

The $\underline{\rho}$ functions and the successive dichotopy procedure discussed in the previous section specifies the pattern recognition block, represented formaly by equation (186). The variable - structure identification scheme is depicted in Fig.14 and the flow-chart of the algorithm is shown in Fig.15.

The process dynamics are described by equation (164) where $\underline{w}(n)$ is the random disturbance. The state $\underline{x}(n)$ is measured by the variable-structure measuring system (167) where the measurement $\underline{z}(n)$ is contaminated by noise $\underline{v}(n)$. The block "estimator" denotes the extended Kalman filter. The optimal measurement structure is specified by classifying the pattern of the estimation variance vector $\underline{\alpha}(n)$. The "pattern recognition" block performs that classification and assigns accordingly the appropriate value of $\underline{c}^*(n)$ such that $\underline{c}^*(n) = \underline{c}^i$ if $\underline{\alpha}(n) \in A^i$.

This completes the synthesis of the on-line identification system.

Illustrative example.

Consider the slab-type nuclear reactor which is represented by the following four-point model [17], based on space discretization where $\underline{X}_n$ is a four-dimensional vector representing the state of the system at the four mech points. The state transition matrix F is given by

$$F = \begin{bmatrix} 0.904 & 0.058 & 0.002 & 0.00039 \\ 0.058 & 0.906 & 0.058 & 0.00185 \\ 0.002 & 0.058 & 0.906 & 0.058 \\ 0.00039 & 0.00185 & 0.058 & 0.904 \end{bmatrix}$$

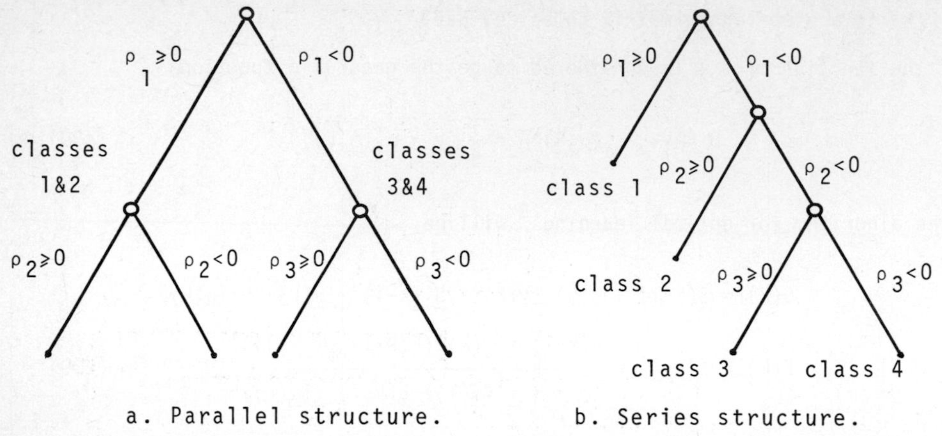

Fig.13 Successive dichotomies in a four-class problem : two possible trees.

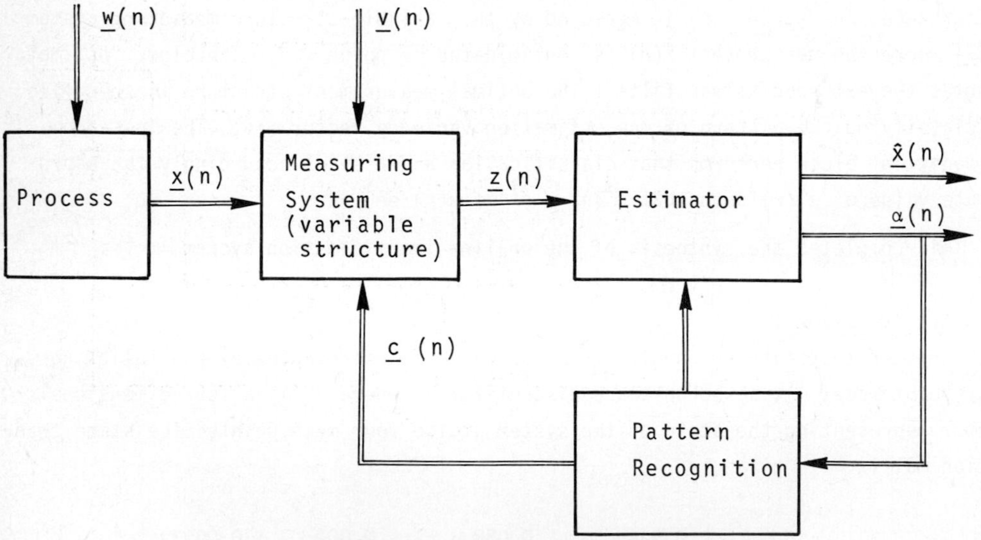

Fig.14 Variable-structure identification scheme.

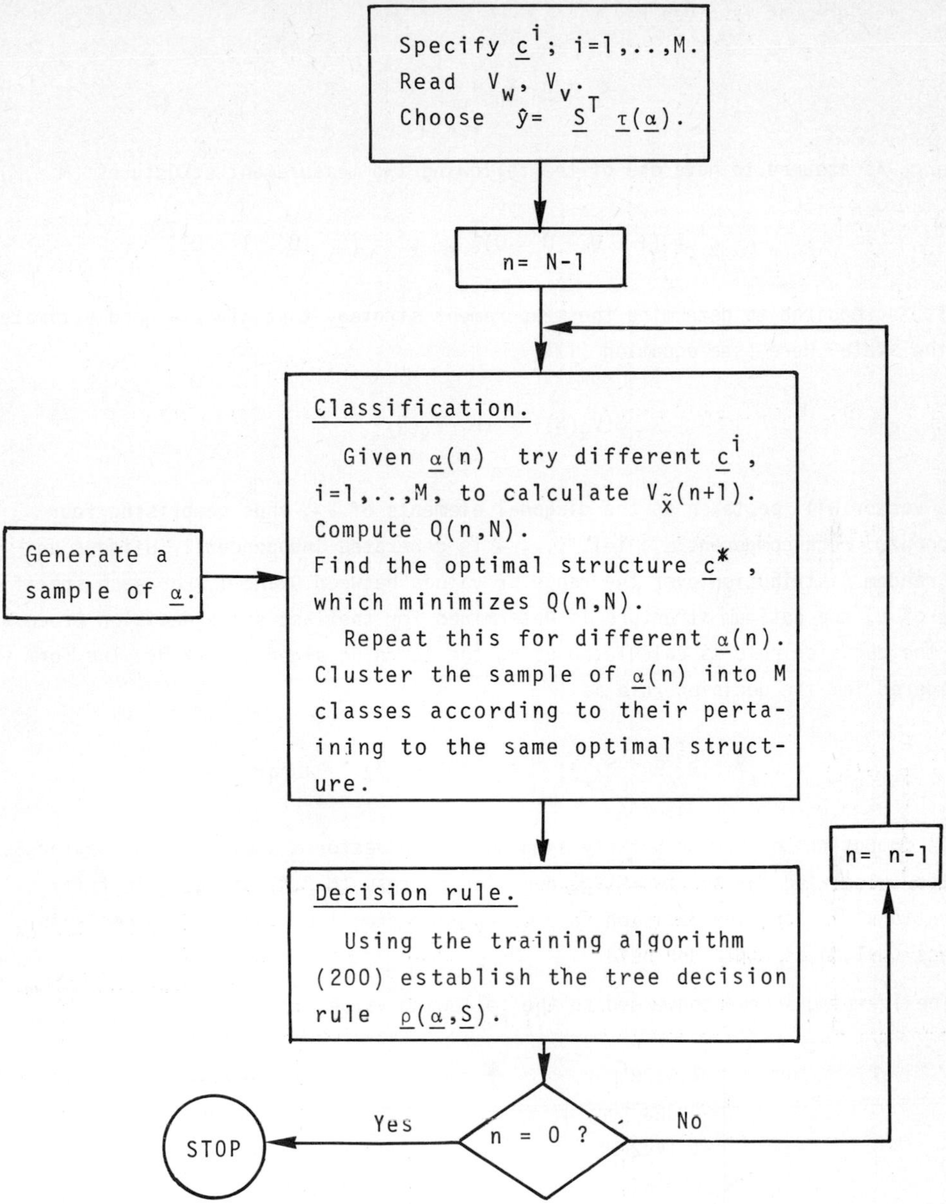

Fig.15 Flow chart of the identification scheme.

The measurement is given by

$$z_n = \underline{c}_n^T \underline{X}_n + v_n$$

where $\underline{c}_n$ is assumed to have one of the following two measurement structures (M = 2):

$$\underline{c}^1 = (1 \quad 0 \quad 0 \quad 0)^T , \quad \underline{c}^2 = (0 \quad 0 \quad 1 \quad 0)^T$$

It is required to determine the measurement strategy that gives a good estimate for the state. Here (see equation (171)

$$\phi\{V_{\tilde{X}}(N)\} = tr \quad V_{\tilde{X}}(N)$$

The $\underline{\alpha}$ vector will be taken as the diagonal elements of $V_{\tilde{X}}$, thus comprising four components. Each component α_i, i=1, ..., 4 is generated independently using a uniform random distribution over the range of values between 0 and 1. For each specific value of $\underline{\alpha}$, the optimum structure is determined for the last stage decision process. Then the decision rule is calculated using the learning algorithm (200). The form considered for the decision rule is

$$\hat{y} = \underline{S}^T \underline{\alpha} = s_1 \alpha_1 + s_2 \alpha_2 + s_3 \alpha_3 + s_4 \alpha_4 .$$

In computing the vector $\underline{S}$ fifty samples of the vector $\underline{\alpha}$ are used. The vector $\underline{S}$ is also calculated for the two-stage decision process (N-2,N) using again fifty generations of $\underline{\alpha}$ and the decision rule calculated from the single-stage decision process (N-1,N) as explained before.

The learning scheme converged to the following values :

$$\underline{S}_{N-1} = \begin{bmatrix} 1.5630 \\ -0.3637 \\ -1.1492 \\ 0.0065 \end{bmatrix} \qquad \underline{S}_{N-2} = \begin{bmatrix} 0.4294 \\ -0.8950 \\ 0.9488 \\ -0.8022 \end{bmatrix}$$

The above decision rules are tested on the same samples of $\underline{\alpha}$ and the misclassification ratio of the optimum structure is 4% for the single-stage and 6% for the two-stages which are considered to be quite satisfactory.

2.16. CONCLUSION.

Pattern recognition consists of two interrelated problems : feature extraction, and classification.

Feature extraction is concerned with extracting, by means of a certain transformation, the important features of the pattern vector with view to classification. The dimension of the feature vector must be as small as possible. Feature extraction consists essentially in extracting what is regarded as valuable information conveyed by a pattern vector. This generally implies that some information conveyed by the pattern vector has to be discarded. There must be some criterion as to what information is to be sought and what to be discarded. If the criterion is to keep the information conveyed by the feature vectors as close as possible to the information conveyed by the respective pattern vectors then the transformation must maximize the entropy in the feature space. That criterion amounts to extracting the features that characterize the attributes common to each pattern class. This is the criterion for intraset feature extraction.

On the other hand, in order to extract the attributes that emphasize the differences between or among pattern classes it is necessary to perform the utmost organization in the feature space (i.e. minimize the entropy) in order to cluster the different populations and hence facilitate separability. This is the criterion for interset feature extraction.

The problem of optimal classification is then stated analytically. In the case of complete a priori information, that problem can be solved by employing the statistical decision algorithms.

If the cost of taking feature measurements is to be included or if the features extracted from the input patterns are sequential in nature then sequential classification methods are to be used.

In that respect Wald's test has been presented. That test becomes practical when considered in its discrete version. To that end finite automata can prove to be a useful tool. Certain form of such automata (automata with linear tactic) is given in some detail.

The Bayes and probabilistic iterative techniques have been presented with view to solving the pattern-recognition problem. Using those techniques different learning algorithms with or without supervision can be obtained. Learning without supervision demands more complex algorithms and takes a longer time than learning with supervision under the same conditions. This agrees with the fact that ignorance must be paid for.

Finally an application of pattern recognition technique to identification of dynamic systems is presented.

COMMENTS

2.1. The division of the pattern recognition problem into extraction and classification problem is rather artificial. Essentially there is no "hard" boundary between both problems. For "perception" aids "decision" inasmuch as "decision" structures what to perceive. An adaptation scheme should be envisaged to adapt the extraction and classification algorithms simultaneously with view to better recognition.

2.2. - 2.5. The terminology of "intraset" and "interset" features is due to Tou and Heydron[1]. They seem to be erroneous in indicating that intraset extraction criterion corresponds to minimizing the entropy, which leads them to an ambiguous result. Also see Tou[2], Young and Calvert[3].

2.8. For Wald's SPRT, see Fu[5]. The automata model is taken from Radyuk and Terpugov[6].

2.10. An interesting discussion about the complexity of the non-supervised Bayes algorithm is also given in Young and Calvert[3] p.83. Approximation of unsupervised Bayes learning are the subject of many researches, see e.g.[18,19].

2.11. The concept of identifiability of finite mixture is originally presented in Teicher[7,8].

2.12. - 2.14. The basic reference is Tsypkin[9]. Accounts of the theory of the iterative probabilistic methods can be found in [10-12].

2.15. That application is due to El-Fattah and Aidarous[13].

REFERENCES

1. J.T. Tou, and R.P. Heydorn, "Some Approaches to Optimum Feature Extraction", in Computer and Information Sciences - II (J.T. Tou, ed.) Academic New York, 1967.

2. J.T. Tou, "Feature Selection for Pattern Recognition Systems", in Methodologies of Pattern Recognition,(S. Watanabe, ed.) New York : Academic, 1972.

3. T.Y. Young, and T.W. Calvert, Classification, Estimation, and Pattern Recognition New York : Elsevier, 1974.

4. W.S. Meisel, Computer-Oriented Approaches to Pattern Recognition. New York : Academic, 1972.

5. K.S. Fu, Sequential Methods in Pattern Recognition and Machine Learning : New York : Academic, 1968.

6. L.E. Radyuk, and A.F. Terpugov, "Effectiveness of Applying Automata with linear Tactic in signal Detection Systems, "Automation and Remote Control", N°4, 1971, pp. 609 - 617.

7. H. Teicher, "Identifiability of Mixtures", Ann. Math. Stat. 32,1961, pp. 244-248.

8. H. Teicher, "Identifiability of Finite Mixtures", Ann. Math. Stat. 34, 1963, pp. 1265 - 1269.

9. Tsypkin Ya. Z. Foundation of the theory of learning systems. Academic Press, 1973, New York.

10. G. Albert, Stochastic Approximation and Non-linear Regrassion. MIT Press, 1967.

11. N.V. Loginov, "Methods of Stochastic Approximation", Automation and Remote Control, 27, N°4, 1966, pp. 706 - 728.

12. D.J. Sakrison, "Stochastic Approximation : A Recursive Method for Solving Regression Problems" in Advan. Communication Systems, 2, 1966.

13. Y.M. El-Fattah, S.E. Aidarous, "A Pattern Recognition Approach for Optimal Measurement Strategies in Dynamic Systems Identification", IFAC sump. on

Identification Tebilisi (USSR), 1976.

14. Aidarous S.E., Gevers M.R., Installe M.J. Int. J. Control, 1975, Vol. 22, 197-213.

15. Athans M. Automatica, 1972, Vol. 8, 397-412.

16. Sage A.P. Estimation and identification. Proc. 5th IFAC Congress, 1972, Paris. (France).

17. Hassan M.A., Ghonaimy M.A.R., Abd El-Shaheed M.A., "A computer algorithm for optimal discrete time state estimation of linear distributed systems", Proc. IFAC Symp. on Control of Distributed Systems, 1971, Banff (Canada).

18. Patrick E.A., J.P. Costello and F.C. Monds, "Decision Directed Estimation of a Two Class Decision Boundary", IEEE Trans. on Computer, vol. C - 19. N°3, pp. 197 - 205, 1970.

19. Makov, U.E., and A.F.M. Smith, "Quasi-Bayes Procedures for Unsupervised Learning" Proc. of the 1976 IEEE Conf. on Decision and Control. Paper WP 4.

20. Rosenblatt, M., "Remarks on some non-parametric estimates of a density functions", Ann-Math. Statist., 27, 832 - 837 (1956).

21. Kullback, S. Information Theory and Statistics. New York : J. Wiley and Sons 1958.

CHAPTER III

SIMULATION - MODELS OF COLLECTIVE BEHAVIOR

"Notre propre intérêt est encore un merveilleux instrument pour nous crever les yeux agréablement".

B. Pascal : Les Pensées.

3.1. INTRODUCTION.

Tsetlin[1] has proposed different norms of behavior of a finite automaton working in a random environment. In that work the environment is assumed to either penalize or reward each action of the automaton according to certain unknown probabilities. The behavior of an automaton is called expedient if the average penalty is less than the value corresponding to choosing all actions with equal probabilities. The behavior is called optimal or ε-optimal according to whether the average penalty is equal or arbitrarily close, respectively, to the minimum value. Krylov and Tsetlin[2] introduced the concept of games between automata and studied in particular Two-Automaton Zero-Sum games.

Stochastic automata with variable structure have been introduced by Varshavskii and Vorntsova[3] to represent learning automata attempting a certain norm of behavior in an unknown random environment. Since the date of that work a respectable number of works has appeared, studying different aspects of learning automata and applying it in simulating very simple norms of behavior (like that introduced by Tsetlin) and also simple automata games (such as Two-Automaton Zero-Sum games). For a survey on the subject we refer to Narendra and Thathachar[4].

The contribution of this chapter is to direct the attention of using learning automata to simulate an important class of problems of collective behavior whose deterministic version has been the subject of recent investigations mainly by Malishevskii and Tenisberg[5-8]. In that class of problems there exists a type of relation in the collective where the behavior of the participants possesses a definite mutual opposition. Such situation can arise for example in economic systems : the case of price regulation in a competitive market[9] ; or in management systems : the problem of resource allocation[10].

In the model introduced in this chapter a collective of interacting stochastic automata is considered. Each automaton has a behavioral tactic directed towards the realization of its own goal, taken to be the minimum of the expected value of a certain penalty function. That function depends explicitly on the automaton strategy and the environment response. The automata interactions arise from the dependence of the environment response on the whole set of strategies used by the collective of automata. That dependence is stochastic and unknown to all the automata. Furthermore, any automaton does not know neither the penalty functions, nor even the number of the other automata. The only available knowledge to each automaton is the realization of its penalty function following the use of a certain strategy.

This model is useful for analysis of some problems of collective behavior in large systems. Its use enables in particular organizing the local behaviors of the different subsystems (constituting a large system) in order to ensure certain desired collective behavior. By organization of local behaviors is generally meant the formation of appropriate criteria (through introduction of penalties or premiums etc.) the provisions of new degrees of freedom, introduction of new links and control levels between the subsystems[11]. In brief, creating an external environment for each subsystem such that the collective behavior of the subsystems - though each pursuing solely its own private interest - is desirable in a definite sense.

3.2. AUTOMATA MODEL I - SUFFICIENT A PRIORI INFORMATION.

As model of collective behavior we consider the following game between N-Learning automata $A^1, A^2, \ldots A^N$, see Fig. 1. A Learning automaton is considered as a stochastic automaton with variable structure as depicted in Fig.2. The automata operate on a discrete time scale $t = 1, 2, \ldots$. The input u^i to a stochastic automaton can only take one of the values -1, 0, +1. The output(.) y^i of the automaton A^i will be assumed to take one of the m^i values $y^i_1, y^i_2, \ldots y^i_{m^i}$, which will be called its stategies. We will say that the automaton A^i uses the jth strategy at time t if $y^i(t) = y^i_j$.

A play $\underline{y}(t)$ carried out at time t will be the name given to the vector $\underline{y}(t) = (y^1(t), y^2(t), \ldots y^N(t))^T$ whose components are the strategies used by the automata $A^1, A^2, \ldots, A^N$ at time t. The outcome $\underline{s}(t+1)$ of a play $\underline{y}(t)$ is the vector $(s^1(t+1), s^2(t+1), \ldots, s^N(t+1))^T$ whose components are the referee or environment responses to the set of automata at time t+1. The environment is completely characterized by the probabilities $p^i(s^i(t+1) / \underline{y}(t))$ of the outcomes $s^i(t+1)$, $i = 1, \ldots, N$, for every play $\underline{y}(t)$. As only stationary environments will be considered, the aforementioned probabilities can simply be written as $p^i(s^i/\underline{y})$.

The probability distribution of the output of the i - th automaton is specified by the vector

(.) a unique deterministic mapping between the automaton states and outputs is assumed.

$$\underline{p}^i = (p_1^i, p_2^i, \ldots., p_{m^i}^i)^T \qquad (i = 1, \ldots., N)$$

$$0 < p_j^i < 1 \qquad \sum_{j=1}^{m^i} p_j^i = 1$$

where p_j^i is the probability that the automaton uses its pure strategy y_j^i. The probability vector $\underline{p}^i$ specifies the mixed strategy of the i - th automaton.

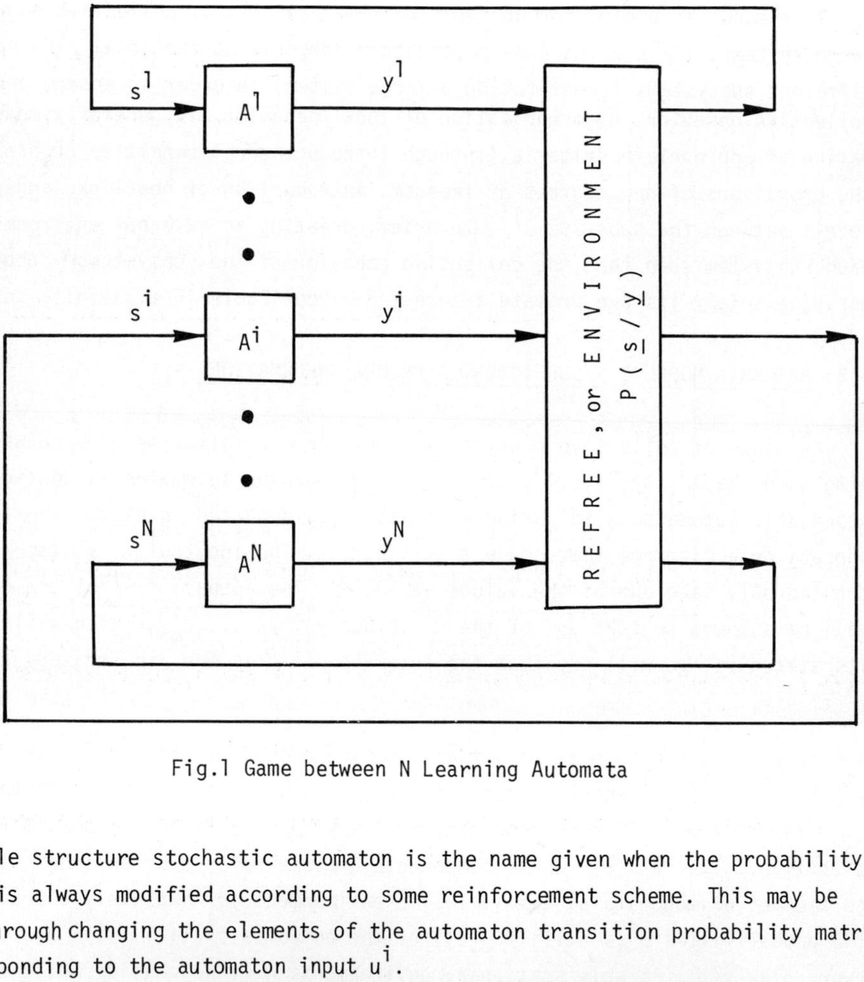

Fig.1 Game between N Learning Automata

Variable structure stochastic automaton is the name given when the probability vector $\underline{p}^i$ is always modified according to some reinforcement scheme. This may be effected through changing the elements of the automaton transition probability matrices corresponding to the automaton input u^i.

The objective of each automaton in the game is to seek the mixed strategy $\underline{p}^i$ that minimizes its average penalty, taken as the expected absolute value of a function F^i that depends on the automaton's strategy y^i and the environment response s^i,

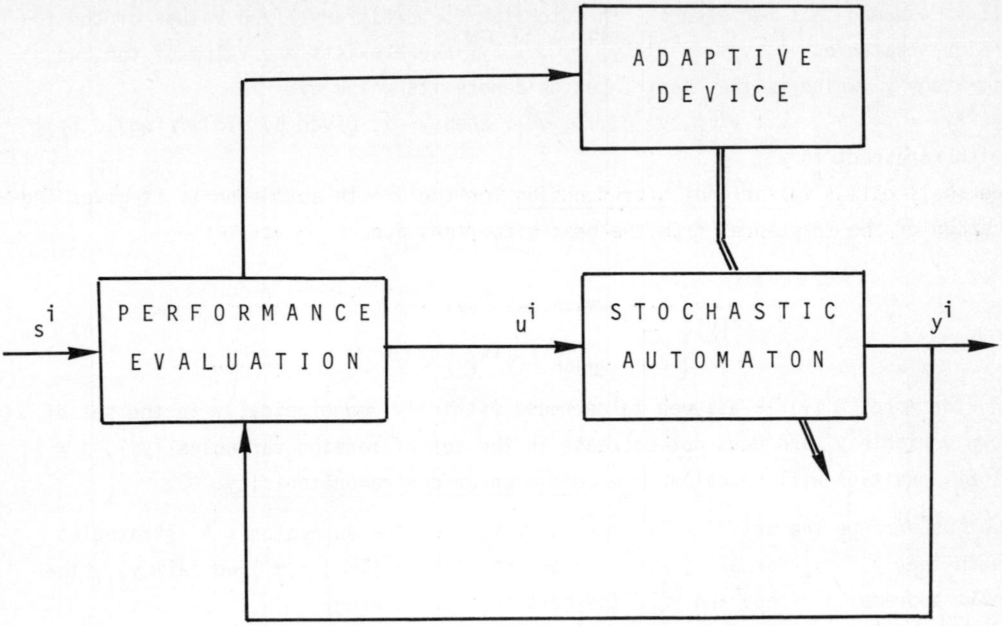

Fig. 2 Learning Automaton

$$Q^i\ (\{\underline{p}^k\}) = E_{s^i,y^i}\ \{|F^i(\Theta^i(y^i,\ s^i))|\}$$

$$= \sum_{j^1\ j^2 ..\ j^N} \int_{-\infty}^{\infty} p^1_{j^1}\ p^2_{j^2}\ ..p^N_{j^N}\ |\ F^i(\Theta^i(y^i_{j^i},s^i))\ |\ dp^i(s^i/\{y^k_{j^k}\}) \qquad (2)$$

where Θ^i is a penalty index which depends on y^i, s^i, and F^i is a function "retaining sign", i.e

$$\text{sign}\ F^i\ (\Theta^i) = \text{sign}\ \Theta^i \qquad (3)$$

The interaction between the automata is obvious from the interdependence of their goals ; notice that Q^i is a function of $\underline{p}^T = (\underline{p}^{1T},\ \underline{p}^{2T}, \ldots\ldots, \underline{p}^{NT})$.

Let us rewrite the average penalty (2) in the form

$$Q^i(\underline{p}) = E_{y^i}\ \{|\delta^i(\underline{y})|\} \qquad (4)$$

where

$$\delta^i(\underline{y}) = E_{s^i}\ \{F^i(\Theta^i(y^i,\ s^i))\}$$

$$= \int_{-\infty}^{\infty} F^i(\Theta^i(y^i,\ s^i))\ d\ p^i(s^i\ /\ \underline{y}) \qquad (5)$$

It is assumed that for each i - th automaton for arbitrary fixed values of the foreign strategies $y^1, \ldots, y^{i-1}, y^{i+1}, \ldots, y^N$ there exists one value of the own strategy y^i which is the "best". Let us denote its value by $y^{i*}(\underline{y}) = y^{i*}(y^1, \ldots, y^{i-1}, y^{i+1}, \ldots, y^N)$. Such y^i is given by minimizing $|\delta^i(\underline{y})|$ with rerespect to y^i.

We shall call $\delta^i(\underline{y})$ an indicator function for the i - th automaton if it gives indication of the "distance" from the best situation, i.e.

$$\delta^i(\underline{y}) \begin{cases} > 0 & \text{when} \quad y^{i*}(\underline{y}) > y^i \\ < 0 & \text{when} \quad y^{i*}(\underline{y}) < y^i \end{cases} \tag{6}$$

Furthermore, $\delta^i(\underline{y})$ is assumed to decrease (strictly) monotonically in the set of own variable y^i and does not decrease in the set of foreign variables $\{y^j\}$, $j \neq i$. That condition will be called the condition of contramonotonicity.

Let us arrange the set $Y^i = \{y_1^i, y_2^i, \ldots, y_{m^i}^i\}$ of the automaton's A^i strategies such that $y_j^i > y_k^i$ for all $j > k$. For any strategy $y_j^i (1 < j < m^i)$ we call y_{j+1}^i the next supremal strategy and y_{j-1}^i the next infimal strategy.

From the contramonotonicity condition the indicator function $\delta^i(\underline{y})$ of the i - th automaton decreases (strictly) with respect to its own variable y^i. This suggests the following operating principle for minimizing the magnitude of δ^i. If $y^i(t) = y_j^i$ resulted in $\delta^i > 0$ use the next supremal strategy, i.e. $y^i(t+1) = y_{j+1}^i$, on the other hand if $\delta^i < 0$ use the next infimal strategy i.e. $y^i(t+1) = y_{j-1}^i$. Otherwise ; on either one of the conditions :

i) $\delta^i = 0$, ii) $y^i(t) = y_{m^i}^i$ and $\delta^i > 0$
iii) $y^i(t) = y_1^i$ and $\delta^i < 0$,

remain in the status quo.

Upon modifying the indicator functions to be in the form

$$\overline{\delta}^i(\underline{y}) = \begin{cases} 0, & \text{if } y^i = y_1^i,\ \delta^i(\underline{y}) < 0 \\ & \text{or } y^i = y_{m^i}^i,\ \delta^i(\underline{y}) > 0, \\ \delta^i(\underline{y}), & \text{otherwise.} \end{cases} \tag{7}$$

The above principles of behavior can be written thus

$$\text{if } y^i(t) = y_j^i \quad \text{then} \quad y^i(t+1) = y_{j+u^i}^i$$

where

$$u^i = \text{sign}\,(\overline{\delta}^i(\underline{y}(t))) \tag{8}$$

In the limit, when the set Y^i tends to the continuous interval $[y_1^i, y_{m^i}^i]$, and the time increment between successive steps tends to zero, the above rule of behavior will approximate the following continuous-time model of collective behavior,

$$\text{sign} \frac{dy^i}{dt} = \text{sign} (\bar{\delta}^i(\underline{y})) \qquad (i = 1,\ldots, N) \quad (9)$$

It was demonstrated by Malishevskii[8] that all nondegenerate trajectories of the process (9) converge to the equilibrium point $\underline{y}^* \varepsilon Y = Y^1 \, x Y^2 \, x \ldots\ldots x Y^N$, if exists, such that

$$\bar{\delta}^i(\underline{y}^*) = 0 \text{ for all } i \quad (10)$$

To realize the above rule of deterministic behavior using stochastic automata, a reinforcement scheme should be divised such that

$$\Pr [y^i(t+1) = y^i_{j+u^i}] > \Pr [y^i(t+1) = y^i_k] \quad \text{for all} \quad k \neq j+u^i \quad (11)$$

An example of such reinforcement scheme may be written thus

$$\text{if} \quad y^i(t) = y^i_{k^i} \text{ then}$$

$$p^i_{k^i+u^i}(t+1) = p^i_{k^i+u^i}(t) + \gamma^i(t+1) \, |\delta^i(\underline{y}(t))|$$

$$p^i_j \,(t+1) = p^i_j(t) - \frac{\gamma^i(t+1)}{m^i - 1} \, |\delta^i(\underline{y}(t))| \quad (12)$$

$$j = 1, \ldots\ldots\ldots\ldots, m^i \qquad j \neq k^i + u^i$$

where $\gamma^i(t+1)$ should satisfy the conditions,

$$\sum_{t=1}^{\infty} \gamma^i(t) = \infty \quad , \quad \sum_{t=1}^{\infty} \gamma^{i2}(t) < \infty, \quad \gamma^i(t) > 0 \quad (13)$$

and subject to the upper bound,

$$\gamma^i(t+1) < \min \left(\frac{1-p^i_{k^i+u^i}(t)}{|\delta^i|} \; ; \; (m^i-1) \frac{\min\limits_{j \neq k^i+u^i} p^i_j(t)}{|\delta^i|}\right), \; \delta^i \neq 0 \quad (14)$$

to guarantee that the probability vector $\underline{p}^i$ satisfies the condition $0 < p^i_j < 1$ for all j.

3.3. AUTOMATA MODEL II - LACK OF A PRIORI INFORMATION.

In the reinforcement scheme (12) complete a priori information is assumed, precisely the joint probability of outcome and play of automata ($p^i(s^i, \underline{y})$) which fully characterizes the external environment.

In the case when such a priori information is unavailable the indicator functions $\delta^i(\underline{y})$, see eqn. (5), cannot be specified explicitly. In such case the scheme (12) is to be replaced by a learning algorithm which provides estimate of the probability vectors $\underline{p}^i(t)$ using the random observations $s^i(t)$. Similar to Kiefer - Wolfowitz stochastic-approximation method[12] we drop the expected-value symbol (with respect to s^i) from the expression of δ^i to get the following reinforcement scheme, provided that

$$y^i(t) = y^i_{k^i}$$

then

$$p^i_{k^i+u^i}(t+1) = p^i_{k^i+u^i}(t) + \gamma^i(t+1) |\, F^i(\Theta^i\,(y^i_{k^i}\; s^i(t+1)))|$$

$$p^i_j(t+1) = p^i_j(t) - \frac{\gamma^i(t+1)}{m^i - 1} \,|\, F^i(\Theta^i(y^i_{k^i},\, s^i(t+1))))| \qquad (15)$$

$$j = 1,\ldots\ldots m^i, \qquad j \neq k^i + u^i$$

where

$$u^i = \text{sign}\, (\bar{\Theta}^i(y^i_{k^i},\, s^i(t+1)) \qquad (16)$$

The sequence $\gamma^i(t+1)$ should also satisfy the condition (12) besides the upper bound condition

$$\gamma^i(t+1) < \min\, \left(\frac{1-p^i_{k^i+u^i}(t)}{|F^i\,(\Theta^i)|},\; (m^i-1)\, \frac{\min\limits_{j\neq k^i+u^i} p^i_j(t)}{|F^i\,(\Theta^i)|}\right), \quad \Theta^i \neq 0 \qquad (17)$$

to guarantee that $p^i_j(t)$ are always between zero and 1.

Similar to (7) the modified penalty index $\bar{\Theta}^i$ is defined as

$$\bar{\Theta}^i(y^i,\, s^i) = \begin{cases} 0, \text{ if } y^i = y^i_1,\; \Theta^i < 0, \\ \text{or } \; y^i = y^i_{m^i},\; \Theta^i > 0, \\ \Theta^i(y^i,\, s^i), \text{ otherwise} \end{cases} \qquad (18)$$

The idea underlying the functioning of a learning automaton in the present model can be stated as follows. At any time step if the automaton action has elicited an environment response for which the penalty index Θ^i is greater than zero then at the next time step the probability of the next supremal action is increased. On the other hand if the penalty index is less than zero then the probability of the next infimal action is increased. If in the case of positive penalty index $y^i_j = y^i_{m^i}$; or in the case of negative penalty index $y^i_j = y^i_1$ then increase the probability of y^i_j. Finally if the penalty index is zero, the automaton remains in the status quo.

3.4. EXISTENCE AND UNIQUENESS OF THE NASH PLAY.

Lemma I.

For contramonotonic indicator functions $\delta^i(\underline{y})$

$$\frac{\Delta y^{i*}(\{y^i\})}{\Delta y^j} > 0 \text{ for all } \Delta y^i \neq 0 \quad (i = 1,\ldots, N) \qquad (19)$$

Proof.

i) Let $\Delta y^j \geq 0$, $j = 1,\ldots, N$, $j \neq i$, and $\Delta y^j > 0$ at least for one j. Let y^{i*} $(\{y^j\})$ denote the best strategy of the i - th automaton for arbitrary fixed values of the foreign strategies $\{y^j\} = \{y^j_k\}$. Let y^{i*} $(\{y^j_k\}) = y^i_\nu$, $1 < \nu < m^i$ and $\delta^i(y^i_\nu, \{y^j_k\}) \geq 0$.

As a consequence of the contramonotonicity assumption

$$\delta^i(y^i_\ell, \{y^j_k + \Delta y^j\}) > \delta^i(y^i_\ell, \{y^j_k\}) > \delta^i(y^i_\nu, \{y^j_k\}) \geq 0$$
$$\text{for all } \ell < \nu$$

Accordingly the optimum y^i for the new $\{y^j\} = \{y^j_k + \Delta y^j\}$ cannot be found in the subset of strategie $(y^i_1, \ldots, y^i_{\nu-1})$ and can only be found in the subset $(y^i_\nu, y^i_{\nu+1}, \ldots, y^i_{m^i})$. Hence $\Delta y^{i*} > 0$. On the other hand if $\delta^i(y^i_\nu, \{y^j_k\}) < 0$ then $\nu = 1$, or else $\delta^i(y^i_{\nu-1}, \{y^j_k\}) > 0$. If $\nu = 1$ then (19) is automatically satisfied. If $\delta^i(y^i_{\nu-1}, \{y^j_k\})$ is positive then $\delta^i(y^i_\ell, \{y^j_k + \Delta y^j\}) > \delta^i(y^i_{\nu-1}, \{y^j_k + \Delta y^j\}) \geq \delta^i(y^i_{\nu-1}, \{y^j_k\}) > 0$ for all $\ell < \nu-1$, whence $\Delta y^{i*} > 0$, and (19) holds.

ii) Let $\Delta y^j \leq 0$, $j = 1,\ldots, N$, $j \neq i$, and $\Delta y^j < 0$ at least for one j. Consider $y^{i*}(\{y^j_k\}) = y^i_\nu$, $1 < \nu < m^i$ and $\delta^i(y^i_\nu, \{y^j_k\}) \leq 0$. As a consequence of the contramonotonicity assumption,

$$\delta^i(y^i_\ell, \{y^j_k + \Delta y^j\}) \leq \delta^i(y^i_\ell, \{y^j_k\}) < \delta^i(y^i_\nu, \{y^j_k\}) \leq 0$$

$$\text{for all } \ell > \nu$$

Hence the optimum y^{i*} for $\{y^j_k + \Delta y^j\}$ can only be found in the subset $\{y^i_1, \ldots, y^i_{\nu-1}\}$. Accordingly $\Delta y^{i*} < 0$ and again (19) is verified. On the other hand if $\delta^i(y^i_\nu, \{y^j_k\}) > 0$ then either $\delta^i(y^i_{\nu+1}, \{y^j_k\}) < 0$ or else $\nu = m^i$. If $\nu = m^i$ then (19) is automatically satisfied. If $\delta^i(y^i_{\nu+1}, \{y^j_k\})$ 0 then $\delta^i(y^i_\ell, \{y^j_k + \Delta y^j\}) \leq \delta^i(y^i_\ell, \{y^j_k\}) < \delta^i(y^i_{\nu+1}, \{y^j_k\}) < 0$ for all $\ell > \nu+1$. Hence $\Delta y^{i*} < 0$, and (19) holds.

Remark :

Lemma I has the following game theoretic interpretation. When strategy y^i of a player A^i has the character of the magnitude of its force, contramonotonicity means that each automaton, in trying to minimize his penalty tries to respond by an increase in his own force to increase of force of the other players.

Theorem I (Existence).

For contramonotonic indicator functions there exists a Nash play $\hat{y}$ such that $\hat{y}^i = y^{i*}(\hat{y})$, (i = 1, 2, ... N).

Proof.

Let $\nu^i_o = 1$ (i = 1, ..., N), and suppose that $y^i_{\nu^i_o}$ is not a Nash play. Consider the set of automata strategies given by

$$y^i_{\nu^i_1} = y^{i*}(\{y^j_{\nu^j_o}\}) \qquad (i = 1, \ldots, N)$$

Suppose that $\{y^i_{\nu^i_1}\}$ is not a Nash play. By virtue of Lemma 1,

$$y^i_{\nu^i_1} \leq y^{i*}(\{y^j\}) \leq y^i_{m^i} \quad \text{whenever} \quad y^j_{\nu^j_1} \leq y^j \leq y^j_{m^j}$$

For all $j \neq i$, (i = 1, ..., N)
i.e. the mappings $y^{i*}(.)$ map the subset of plays

$$Y^{(1)} = \{\underline{y} : y^i_{\nu^i_1} \leq y^i \leq y^i_{m^i} \quad (i = 1, \ldots N)\}$$

into itself. Thus a Nash play must exist in $Y^{(1)}$. Let us then consider the strategies,

$$y^i_{\nu^i_2} = y^{i*}(\{y^j_{\nu^j_1}\}) \qquad (i = 1, \ldots, N)$$

If $\{y^i_{\nu^i_2}\}$ again is not a Nash play, then a Nash play must be in the subset,

$$Y^{(2)} = \{\underline{y} : y^i_{\nu^i_2} \le y^i \le y^i_{m^i} \qquad (i = 1, \ldots, N)\}$$

and the mappings $y^{i*}(.)$ map $Y^{(2)}$ into itself. By successive application of the above procedure it follows that if $\{y^i_{\nu^i_s}\}$ is not a Nash play then the candidate for that play must be in the subset of plays

$$Y^{(s)} = \{\underline{y} : y^i_{\nu^i_s} < y^i < y^i_{m^i} \qquad (i = 1, \ldots, N)\}$$

By Lemma I it is clear that

$$y^i_{\nu^i_0} \le y^i_{\nu^i_1} \le \ldots\ldots \le y^i_{\nu^i_s} \le \ldots\ldots \le y^i_{m^i}$$

For all i. Since $y^{i*}(.)$ map the subset $Y^{(s)}$ into itself, it is obvious that in the limit - unless $y^i_{\nu^i_{\hat{s}}}$ is a Nash play for some $\hat{s}$ - the subset of candidate Nash plays will degenerate to the "boundary" point $\{y^i_{m^i}\}$. The proof is complete.

Remark :

Two situations where a Nash play is obvious are,

i) $\delta^i(\underline{y})$ is always positive for all i and $\underline{y}$. The Nash play will be the boundary point $\{y^i_{m^i}\}$.

ii) $\delta^i(\underline{y})$ is always negative for all i and $\underline{y}$. The Nash play will be the boundary point $\{y^i_1\}$.

Henceforth we shall be considering collections of functions-indicators $\delta^1(\underline{y}), \ldots$ $\ldots\ldots \delta^N(\underline{y})$, satisfying the condition formulated below. In the adopted formulations, $\underline{y}$ and $\underline{y} + \Delta\underline{y}$ are arbitrary points of $Y = Y^1 \times Y^2 \times \ldots \times Y^N$, while $\Delta\delta^i(\underline{y}) = \delta^i(\underline{y} + \Delta\underline{y}) - \delta^i(\underline{y})$.

CONDITION I :

Let $\Delta\underline{y} \ne \underline{0}$. We so partition the set of subscripts $I=\{1,2,..,N\}$ into three subsets $I_>$, $I_=$, $I_<$; such that $i\varepsilon I_> \leftrightarrow \Delta y^i>0$, $i\varepsilon I_= \leftrightarrow \Delta y^i=0$, $i\varepsilon I_< \leftrightarrow \Delta y^i< 0$.
The following inequality is then assumed to hold :

$$\sum_{i \,\varepsilon\, I_>} \Delta\delta^i(\underline{y}) - \sum_{i \,\varepsilon\, I_<} \Delta\delta^i(\underline{y}) + \sum_{i \,\varepsilon\, I_=} \Delta\delta^i(\underline{y}) < 0$$

$$(\Delta\underline{y} \ne 0) \text{ for any } \underline{y} \,\varepsilon\, Y \qquad (20)$$

We readily see that the above condition presumes, in particular, monotonic decrease of functions $\delta^i(\underline{y})$ with $y^i (i = 1, ..., N)$. That condition can be considered as a variant of the concept of a monotonically decreasing function, generalized to the case of vector function $\underline{\delta}(\underline{y})$, i.e. , the collection of N functions of N variables $(\delta^1(y^1, ..., y^N), ..., \delta^N(y^1, ..., y^N))$.

One can also see the appearance of the boundedness of the inter-automata influences if one provides the following interpretation : the interaction among several goal-oriented automata constitute competition among the automata users for some resource which is necessary to them. Let y^i be the magnitude of the effort of the i - th automaton to acquire the resource, while $\delta^i(\underline{y})$ is the magnitude of the deficit of the resource (if $\delta^i < 0$ then $|\delta^i|$ is the magnitude of the excess) in terms of the i - th automaton in the play $\underline{y}$. Then the monotonicity of $\delta^i(\underline{y})$ with respect to y^i, assumed by the condition I, means that there is the possibility of self-regulation by each automaton individually, since there is guaranteed a decrease of the resource available as its own efforts increase (and conversely).

In the given interpretation, let group $I_>$ consist of the automata increasing on (in any case not decreasing) their efforts to acquire the resource while group $I_<$, on the contrary, consist of the automata decreasing (not increasing) their efforts. Then, according to condition I, the total deficit $\Sigma\delta^i$ for the first group is decreased in comparison with the second group. This can be treated as a presupposition for the possibility of the group self-regulation of such automata.

In brief condition I reflects the retention of the capability of self-regulation in a system of automata whose mutual influences are bounded.

CONDITION 2 :

For any i, and for any arbitrary fixed foreign strategies, $\delta^i(\underline{y})$ as a function of the own strategy y^i does not assume two consecutive values which are equal in magnitude, i.e.

$$|\delta^i(y^i, \ldots, y^{i-1}, y^i_j, y^{i+1}, \ldots, y^N)| > |\delta^i(y^1, \ldots y^{i-1}, y^i_{j+1}, y^{i+1}, \ldots, y^N)|$$

$$j = 1, 2, \ldots, m^i - 1 \qquad (21)$$

$$(i = 1, \ldots, N)$$

Condition 2 is necessary to guarantee that for any arbitrary fixed foreign strategies there exists only one best own strategy,

$$y^{i*}(\underline{y}) , \quad (i = 1, \ldots, N)$$

We now investigate the uniqueness of the Nash play. Let us introduce the auxiliary function

$$\Phi(\underline{y}) = \sum_{i=1}^{N} |\delta^i(\underline{y})| \qquad (22)$$

Lemma 2.

Let condition I hold. Let $\delta^i(\underline{y})\ \Delta y^i \leqslant 0$ $(i = 1, \ldots, N)$ and $\Delta\underline{y} \neq 0$. Then $\Delta\Phi(\underline{y}) > 0$.

Proof.

We write the evident relationship

$$\Delta|\delta^i(\underline{y})| = |\delta^i(\underline{y} + \Delta\underline{y})| - |\delta^i(\underline{y})| \geqslant - |\delta^i(\underline{y} + \Delta\underline{y}) - \delta^i(\underline{y})| = - |\Delta\delta^i(\underline{y})| \tag{23}$$

We use (23) to estimate those terms in $\Delta\Phi(\underline{y}) = \sum_i \Delta|\delta^i(\underline{y})|$ corresponding to $\Delta y^i = 0$. For those i for which $\Delta y^i \neq 0$ we shall provide another estimate. First, taking into account that $|\text{sign}\ \Delta y^i| = 1$, we have

$$|\delta^i(\underline{y} + \Delta\underline{y})| \geqslant - \delta^i(\underline{y} + \Delta\underline{y}) \text{sign}\ \Delta y^i \tag{24}$$

Second, in view of the conditions of the lemma, when $\Delta y^i \neq 0$ and $\delta^i(\underline{y}) \neq 0$ we have $\text{sign}\ \delta^i(\underline{y}) = - \text{sign}\ \Delta y^i$ so that, when $\Delta y^i \neq 0$

$$|\delta^i(\underline{y})| = \delta^i(\underline{y})\ \text{sign}\ \delta^i(\underline{y}) = - \delta^i(\underline{y})\ \text{sign}\ \Delta y^i \tag{25}$$

From (24) and (25) we find

$$\Delta|\delta^i(\underline{y})| > - \Delta\delta^i(\underline{y})\ \text{sign}\ \Delta y^i, \text{ if } \Delta y^i \neq 0 \tag{26}$$

From (23) and (26) we find

$$\Delta\Phi(\underline{y}) = \sum_{i=1}^{N} \Delta|\delta^i(\underline{y})| = \sum_{i:\Delta y^i \neq 0} \Delta|\delta^i(\underline{y})| + \sum_{i:\Delta y^i = 0} \Delta|\delta^i(y) >$$

$$- \left(\sum_{i:\Delta y^i \neq 0} \Delta\delta^i(\underline{y})\ \ \text{sign}\ \Delta y^i + \sum_{i:\Delta y^i = 0} |\Delta\delta^i(\underline{y})|\right) \tag{27}$$

so that by virtue of condition I, $\Delta\Phi(\underline{y}) > 0$.

Theorem 2 (Uniqueness).

For $\underline{y}$ to be a Nash play it is necessary and sufficient that $\underline{y}$ be a minimum point of the function $\Phi(\underline{y})$. $\underline{y}$ is unique.

Proof.

Sufficiency. If $\underline{y}^*$ is a Nash play then there exists three possibilities :

i) $\delta^i(\underline{y}^*) = 0$ for all i,

ii) $\delta^i(\underline{y}^*) < 0$ for some i,

iii) $\delta^i(\underline{y}^*) > 0$ for some i.

In the first case it is trivial to see that $\delta^i(\underline{y}^*)\Delta y^i = 0$ for all i.
In the second case, with due regard to the monotonicity of $\delta^i(\underline{y})$, it is obvious that the next infimal strategy to y^{i*} will correspond to a positive δ^i. Let $y^{i*} = y^i_\nu$. Then $\Delta y^i\ \delta^i(\underline{y}^*) < 0$, $\Delta y^i > 0$, $\Delta y^i\ \delta^i(y^1, \ldots, y^{i-1}, y^i_{-1}, y^{i+1}, \ldots, y^N) < 0$, $y^i < 0$.
In the third case, by similar reasoning, it is obvious that the next supremal strategy to y^i will correspond to a negative δ^i. Then $\Delta y^i\ \delta^i(\underline{y}^*) < 0$, $\Delta y^i < 0$, and $\Delta y^i\ \delta^i(y^1, \ldots, y^{i-1}, y^i_{+1}, y^{i+1}, \ldots, y^N) < 0$, $y^i > 0$.
It follows then that in every case $\Delta y^i\ \delta^i(\underline{y}^*) \leqslant 0$ for all i, and consequently $\Delta\Phi(\underline{y}) > 0$. Hence $\underline{y}^*$ is the minimum point of the function Φ on Y and is unique as well, by virtue of condition 2.

Necessity.

Let $\underline{y}^* = \{y^i_{\nu^i}\}$ be the minimum point of Φ. Assume that $\underline{y}^*$ is not a Nash play. That means with due regard to the monotonicity property that

$$\text{sign}\ \delta^i(y^i_{\nu^i-1}) = \text{sign}\ \delta^i(y^i_{\nu^i}) \text{ if } \delta^i(\underline{y}^*) \text{ is negative or}$$
$$\text{sign}\ \delta^i(y^i_{\nu^i+1}) = \text{sign}\ \delta^i(y^i_{\nu^i}) \text{ if } \delta^i(\underline{y}^*) \text{ is positive.}$$

Let us then consider the point $\underline{y}$ obtained by replacing the components of y^i_ν of $\underline{y}^*$ by either $y^i_{\nu-1}$ or $y^i_{\nu+1}$ depending on whether δ^i is negative or positive, respectively.
Then we get the inequality $\delta^i(\underline{y})\ (y^{i*} - y^i) \leqslant 0$ for all i. Hence $\Phi(\underline{y}^*) > \Phi(\underline{y})$ which is contrary to the assumption that $\underline{y}^*$ is the minimal point of Φ. Hence $\underline{y}^*$ must be the Nash play. The proof is complete.

3.5. CONVERGENCE THEOREM.

Instead of the reinforcement scheme (15) we present a more generalized version which is called projectional algorithm (in contradistinction with the projectionless version (15)). The advantage of the projectional algorithm is to get rid of the constraint (17) which has to be checked at each time instant. The algorithm is given thus ,

$$\text{if } y^i(t) = y^i_{k^i}, \text{ then}$$

$$p^i_{k^i+u^i}(t+1) = \Omega_{S_{\varepsilon^i(t+1)}}\{p^i_{k^i+u^i}(t) + \gamma^i(t+1)|F^i(\Theta^i(y^i_{k^i}, s^i(t+1)))|\} \tag{28}$$

$$p^i_j(t+1) = \Omega_{S_{\varepsilon^i(t+1)}}\{p^i_j(t) - \frac{\gamma^i(t+1)}{m^i - 1}|F^i(\Theta^i(y^i_{k^i}, s^i(t+1)))|\}$$

$$j \neq k^i + u^i \qquad (i = 1, \ldots\ldots, N)$$

where u^i is defined as in (16). Ω_{S_ε} denotes the projection operator into the simplex $S_\varepsilon = \{\underline{p}^i : p_j^i > \varepsilon^i, \sum_{j=1}^{m^i} p_j^i = 1\}$.

Theorem 3.

The automata play $\underline{y}(t)$ converges in probability to the Nash play, i.e. $\underline{y}(t) \xrightarrow{p} \hat{\underline{y}}$, if the conditions,

$$\gamma^i(t) > 0 \qquad \sum_{t=1}^{\infty} \gamma^i(t) = \infty, \quad \sum_{t=1}^{\infty} \gamma^{i2}(t) < \infty \tag{29}$$

$$\varepsilon^i(t) > 0 \qquad \sum_{t=1}^{\infty} \gamma^i(t)\, \varepsilon^i(t) < \infty, \ \varepsilon^i(t) \xrightarrow[t \to \infty]{} 0$$

$$(i = 1, \ldots.., N)$$

hold for the reinforcement scheme (28).

For the proof of the above theorem we shall make use of the following theorem on the convergence of almost super-martingales which has been proved by Robbins and Siegmund[13].

Theorem 4.

Let (Ω, F, P) be, a probability space and $F_1 \subset F_2 \subset \ldots.$ be a sequence of sub σ- fields of F. Let U_t, β_t, ξ_t and ζ_t, $t = 1, 2, \ldots.$, be non negative F_t - measurable random variables such that

$$E(U_{t+1}/F_t) \leqslant (1 + \beta_t)\, U_t + \xi_t - \zeta_t, \qquad t = 1, 2 \ldots.$$

Then on the set $\{\sum_t \beta_t < \infty, \sum_t \xi_t < \infty\}$ U_t converges a.s. to a random variable and $\sum_t \zeta_t < \infty$ a.s.

Proof of Theorem 3.

Let

$$V(\underline{p}(t)) = \sum_{i=1}^{N} \sum_{j=1}^{m^i} (p_j^i(t) - p_j^{i*})^2 \tag{30}$$

In view of the reinforcement scheme (28), V can be rewritten as

$$V(\underline{p}(t)) = \sum_{i=1}^{N} [\Omega_{S_{\varepsilon(t)}}(p^i_{k^i+u^i}(t-1) + \gamma^i(t) |F^i(\theta^i(y^i_{k^i}, s^i))|) - p^{i*}_{k^i+u^i}]^2 +$$

$$\sum_{i=1} \sum_{j \neq k^i+u^i} [\Omega_{S_{\varepsilon(t)}}(p_j^i(t-1) - \frac{\gamma^i(t)}{m^i - 1} |F^i(\theta^i(y^i_{k^i}, s^i))|) - p_j^{i*}]^2 .$$

Using the contraction property of the projection operator, we get,

$$V(\underline{p}(t)) \leq \sum_{i=1}^{N}\sum_{j=1}^{m^i}(p_j^i(t-1)-p_j^{i*})^2 + \sum_{i=1}^{N}\gamma^{i2}(t)\,|F^i(\Theta^i(y^i_{k^i},s^i))|^2 +$$

$$+ \sum_{i=1}^{N}\sum_{j\neq k^i+u^i}\frac{\gamma^{i2}(t)}{(m^i-1)^2}\,|F^i(\Theta^i(y^i_{k^i},s^i))|^2 +$$

$$+ \sum_{i=1}^{N} 2\,\gamma^i(t)(\,p^i_{k^i+u^i}(t-1)-p^{i*}_{k^i+u^i})\,|F^i(\Theta^i(y^i_{k^i},s^i))|$$

$$- \sum_{i=1}^{N}\sum_{j\neq k^i+u^i} 2\,\frac{\gamma^i(t)}{m^i-1}(\,p_j^i(t-1)-p_j^{i*})\,|F^i(\Theta^i(y^i_{k^i},s^i))|$$

$$= V(\underline{p}(t-1)) + \sum_{i=1}^{N}\frac{m^i}{m^i-1}\,\gamma^{i2}(t)\,|F^i(\Theta^i(y^i_{k^i},s^i))|^2 +$$

$$+ \sum_{i=1}^{N}\frac{2\,m^i}{m^i-1}\,\gamma^i(t)\,(\,p^i_{k^i+u^i}(t-1)-p^{i*}_{k^i+u^i})|\,F^i(\Theta^i(y^i_{k^i},s^i))|$$

$$- \sum_{i=1}^{N}\frac{2}{m^i-1}\,\gamma^i(t)\,\underbrace{\sum_{j=1}^{m^i}(p_j^i(t-1)-p_j^{i*})}_{=\,0}\,|F^i(\Theta^i(y^i_{k^i},s^i))|$$

Averaging both sides with respect to the random variable s^i and the random indices k^i, u^i for fixed $\underline{p}(t-1)$, we obtain :

$$E\,V(\underline{p}(t))/\underline{p}(t-1) < V(\underline{p}(t-1)) +$$

$$+ \sum_{i=1}^{N}\gamma^{i2}(t)\,\frac{m^i}{m^i-1}\sum_{k^1,k^2,\ldots,k^N} E\{|F^i(\Theta^i(y^i_{k^i},s^i))|^2/\{k^j\}\}p^1_{k^1}(t-1)\ldots p^N_{k^N}(t-1)$$

$$+ \sum_{i=1}^{N}\gamma^i(t)\,\frac{2\,m^i}{m^i-1}\Bigg[\sum_{k^i=1}^{m^i-1}(p^i_{k^i+1}(t-1)-p^{i*}_{k^i+1})E\{[F^i(\Theta^i(y^i,s^i))]^+\}\,p^i_{k^i}(t-1) +$$

$$+ \;(p^i_{m^i}(t-1)-p^{i*}_{m^i})\;E\{[F^i(\Theta^i(y^i,s^i))]^+\}\,p^i_{m^i}(t-1) +$$

$$+ \sum_{k^i=2}^{m^i}(p^i_{k^i-1}(t-1)-p^{i*}_{k^i-1})E\{[F^i(\Theta^i(y^i,s^i))]^-\}\,p^i_{k^i}(t-1) +$$

$$+ (p_1^i(t-1)-p_1^{i*})\;E\{[F^i(\Theta^i(y^i,s^i))]^-\}\,p_1^i(t-1)\Bigg] \tag{31}$$

where

$$[x]^+ = \begin{cases} x, & \text{if } x > 0 \\ 0, & \text{if } x \leq 0 \end{cases} \tag{32}$$

and $[x]^- = [-x]^+$.

The third term in (31) is indeed equivalent to,

$$E\{\sum_{\substack{i=1\\ \Delta y^i>0}}^{N} \gamma^i(t)\frac{2\,m^i}{m^i-1}\,\Delta\delta^i(\underline{y})+\sum_{\substack{i=1\\ \Delta y^i=0}}^{N} \gamma^i(t)\frac{2\,m^i}{m^i-1}\,\Delta\delta^i(\underline{y})-\sum_{\substack{i=1\\ \Delta y^i<0}}^{N} \gamma^i(t)\frac{2\,m^i}{m^i-1}\,\Delta\delta^i(\underline{y})\} \tag{33}$$

which is negative for any $\Delta\underline{y} \neq \underline{0}$ and all $\underline{y}\,\varepsilon\,Y$, by virtue of condition I (cf. Sec. 3.4).

Note that $E\{\Delta\underline{y}\} = \underline{0}$ if and only if $\underline{p}^i(t-1) = \underline{p}^{i*}$ for all i. Therefore, we can assign positive constants $B^1, B^2,..,B^N$ such that the third term in (31) is expressed as,

$$-\sum_{i=1}^{N} \gamma^i(t)\; B^i \,\|\, \underline{p}^i(t-1) - \underline{p}^{i*} \,\| \tag{34}$$

Let us introduce the upper bounds,

$$0 < \sum_{k^1,k^2,..,k^N} \frac{m^i}{m^i-1}\; E\{|F^i(\Theta^i(y^i_{k^i},s^i))|^2/\{k^j\}\}p^1_{k^1}\; p^2_{k^2}..p^N_{k^N}$$

$$\max_{k^1,k^2,..,k^N} \; E\{|F^i(\Theta^i(y^i_{k^i},s^i))|^2/\{k^j\}\} = A^i < \infty \quad , \; i=1,..,N. \tag{35}$$

Hence inequality (31) defines the almost supermartingale,

$$E\,\{V(\underline{p}(t+1)/\underline{p}(t)\}<V(\underline{p}(t))+\sum_{i=1}^{N} A^i\;\gamma^{i2}(t) - \sum_{i=1}^{N} \gamma^i(t)\;B^i \,\|\,\underline{p}^i(t)\; -p^{i*}\,\| \tag{36}$$

Applying the aforementioned theorem of Robbins and Siegmund, we conclude that $V(\underline{p}\,(t))$ converges a.s. to a random variable and

$$\sum_t \gamma^i(t)\|p^i(t)- p^i\| < \infty\;, \quad i = 1,\; \ldots,\; N \tag{37}$$

on the condition

$$\sum_t \gamma^{i2}(t) < \infty \qquad i= 1,\; \ldots,\; N \tag{38}$$

Noting that

$$||\underline{p}^i(t) - \underline{p}^{i*}|| < ||\underline{p}^i(t) - \hat{\underline{p}}^i_\varepsilon(t)|| + ||\hat{\underline{p}}^i_\varepsilon(t) - \underline{p}^{i*}||$$

where $\hat{\underline{p}}^i_\varepsilon(t)$ is the point on $S_{\varepsilon(t)}$ which is closest to $\underline{p}^{i*}$, we write

$$B^i \| \underline{p}(t) - \underline{p}^{i*} \| < B^i ||\underline{p}^i(t) - \hat{\underline{p}}^i_\varepsilon(t)|| + B^i \varepsilon^i(t) \tag{39}$$

According to (37) and (39) the sequence $\{\varepsilon^i(t)\}$ must guarantee the convergence of the series $\gamma^i(t)\,\varepsilon^i(t)$, i.e.

$$\sum_{t=1}^{\infty} \gamma^i(t)\,\varepsilon^i(t) < \infty \tag{40}$$

With due regard to (37), (39), (40) as well as the fact that $\{\gamma^i(t)\}$ is divergent,

$$\sum_{t=1}^{\infty} \gamma^i(t) = \infty$$

We get $\underline{p}^i(t) \xrightarrow{P} \hat{\underline{p}}^i_\varepsilon$. Since $\hat{\underline{p}}^i_\varepsilon \longrightarrow \underline{p}^{i*}$ as $t \longrightarrow \infty$ as a consequence of the fact that

$$\varepsilon^i(t) \xrightarrow[t \to \infty]{} 0 \tag{41}$$

Then $\{\underline{p}^i(t)\}$ converge in probability to $\{\underline{p}^{i*}\}$, $i = 1, \ldots\ldots, N$. The proof is complete.

3.6. ENVIRONMENT MODEL.

As said before the environment is completely characterized by the probabilities $\{\underline{p}^i(s^i, \underline{y})\}$ of the outcomes $s^i(t+1)$ for every play $\underline{y}(t)$. These probabilities fully specify the automata interactions.

In the following we present two different models of the environment, namely the "pairwise comparison" and the "proportional utility".

3.6.1. Pairwise comparison.

Let the environment be constituted of ν elements $j = 1, \ldots, \nu$. The j - th element finds out the strategies y^i and y^k of two randomly chosen (with equal probabilities) automaton ; the i - th and the k - th ($i, k = 1, \ldots, N$). The j - th element then responds in a probabilistic manner to only one of the chosen pair of automata : say with probability $p^j(y^i, y^k)$ to the i - th and with probability $p^j(y^k, y^i) = 1 - p^j(y^i, y^k)$ to the k - th.

We shall assume $p^j(y^i, y^k) = \psi(\rho^j(y^i, y^k) = 1/2 + \mu(\rho^j(y^i, y^k))$ where $\rho^j(y^i, y^k)$ is a certain utility index for the j - th element, and $\mu(x)$ is a monotonically increasing odd function $\mu(+\infty) = -\mu(-\infty) = 1/2$.

The considered form for the probability of a response from an environment element to an automaton agrees with the natural assumption that the probability increases as the utility of the element increases and vice versa. The dependence of the elements' utility on the automata strategies y^i, y^k($i, k = 1, \ldots, N$) sets the competition and consequently stimulates certain objectives for the automata. For this model of pairwise comparison the competition comes from the fact that for any automaton, say the i - th, another automaton e.g. the k - th, while seeking its own goal, may minimize the utility of an environment element response to the i - th automaton. It is therefore conspicuous that the utility p^j has to be a function of the difference between y^i and y^k. The sign of that difference determines on which side the strategy y^i is to be manipulated by the i - th automaton.

Since the probability of choosing two automata out of N ones equals 2/N(N-1), it is clear that the total probability of agreement between a j - th element and an i - th automaton is equal to

$$p^{ji}(\underline{y}) = \frac{2}{N(N-1)} \sum_{\substack{k=1 \\ k \neq i}}^{N} \psi(\rho^j(y^i, y^k)) \qquad (42)$$

Notice that

$$0 < p^{ji} < 1, \quad \sum_{i=1}^{N} p^{ji} = 1 \tag{43}$$

Let the response function of the j - th element to the i - th automaton is given by

$$s^i = \begin{cases} \omega(y^i) & , \quad y^i \leq \xi \\ 0 & , \quad \text{otherwise} \end{cases} \tag{44}$$

where ω is some positive piecewise continuous function, and ξ is a certain threshold value.

Let $p_0^i(\{y^k\})$ denote the probability of no response, i.e. $0 \leqslant s^i < \omega(y^i)$. Then the probability of one response i.e. $(\omega(y^i) \leqslant s^i < 2\omega(y^i))$ by one element out of ν is equal to

$$p^i(\omega(y^i) \leqslant s^i < 2\omega(y^i) | \{y^k\}) = p_0^i(\{y^k\}) + \max_{1 \leqslant j_1 \leqslant \nu} p^{j_1 i}(\{y^k\}) \cdot sg\,(\xi - y^i) \tag{45}$$

where

$$sg\,(x) = \begin{cases} 1 & , \quad x \geqslant 0 \\ 0 & , \quad x < 0 \end{cases} \tag{46}$$

The function sg (x) is introduced to respect the boundedness condition (44).

Also, the probability of two responses (i.e. $2\omega(y^i) \leqslant s^i < 3\omega(y^i)$) by two elements out of ν is equal to :

$$p^i(2\omega(y^i) \leqslant s^i < 3\omega(y^i)/\{y^k\}) = p_0^i(\{y^k\}) + \frac{1}{2}\,[p^{j_1 i}(\{y^k\}) + \max_{\substack{1 \leqslant j_2 \leqslant \nu \\ j_2 \neq j_1}} p^{j_2 i}(\, y^k\,)\, sg(\xi - y^i) \tag{47}$$

Analogously, the probability of n responses (i.e. $n\omega(y^i) \leqslant s^i < (n+1)\omega(y^i)$) by n elements out of ν is equal to

$$p^i(n\omega(y^i) < s^i < (n+1)\omega(y^i)/\, y^k\,) = p_0^i(\{y^k\}) + \frac{1}{n}\Big[\sum_{\ell=1}^{N-1} p^{j_\ell i}(\, y^k\,) + \max_{\substack{1 \leqslant j_n \leqslant \nu \\ j_n \neq j_1, j_2, \ldots, j_{n-1}}} p^{j_n i}(\{y^k\})\, sg(\xi - y^i)\Big] \tag{48}$$

The probability of no response $p_0^i(\{y^k\})$ is such as to fulfil the normalization condition

$$\Sigma_{s^i} \, p^i(s^i/\underline{y}) = 1 \qquad (49)$$

3.6.2. Proportional utility.

In this model each element of the environment responds to the automata with probabilities proportionable to the utilities of their strategies. The probability of a response from an element increases as the utility of an automaton strategy increases and becomes maximum for maximum utility. Hence the probability that the j - th element responds to the i - th automaton can be expressed thus,

$$p^{ji}(\underline{y}) = \psi(\rho^j(y^i)) / \sum_{k=1}^{N} \psi(\rho^j(y^k)), \qquad (50)$$

$$j = 1, \ldots\ldots, \nu \; ; \qquad i = 1, \ldots\ldots, N$$

where $\psi(.)$ is a positive non-decreasing function, as described beforehand, and $\rho^j(y^i)$ is the utility of the i - th automaton strategy y^i for the j - th element. Notice that the probabilities $p^{ji}(\underline{y})$ satisfy condition (43).

If the utility of an element decreases as y^i increases, then the probability (50) will have the necessary property of contramonotonicity, i.e. it decreases monotonically with respect to the output y^i of the i - th automaton, and increases monotonically with respect to the set of other automata outputs $y^k (k \neq i)$.

Eqs. (44) - (48) again complete the mathematical description of this environment model.

3.7. MARKET PRICE FORMATION.

Consider N sellers in a market trading in one specific commodity. Each i - th seller (i = 1, ..., N) is assumed to be supplied by a constant q^i units of that commodity per time increment (the interval between any two successive time steps). The strategy of any i - th seller y^i represents the price he specifies for his commodity. Let the i - th seller receives a demand s^i in monetary units for buying his commodity at the specified price y^i. The penalty index for the i - th seller is simply the mis-match between the demand and supply in monetary units, i.e.

$$\theta^i = s^i - q^i y^i \quad , \quad (i = 1, \ldots, N) \qquad (51)$$

The consequence of that mis-match may be interpreted differently by the sellers ; each according to his psychological type. That interpretation is embodied in the weighting function $F^i(.)$ of an i - th seller which may be considered in the following form :

$$F^i(\Theta^i) = a^i(\exp(b^i\Theta^i) - 1) + d^i\Theta^i,$$

$$(i = 1, \ldots, N) \tag{52}$$

The constants a^i, b^i, and d^i simulate the psychological type of the i - th seller as follows :

$$\begin{array}{lll} \text{Cautions type} & : & a^i, b^i < 0 \;,\; d^i = 0 \\ \text{Objective type} & : & a^i, b^i = 0 \;,\; d^i > 0 \\ \text{Hazardous type} & : & a^i, b^i > 0 \;,\; d^i = 0 \end{array} \tag{53}$$

The nonlinearity of the weighting function F^i for cautious or hazardous seller indicates the lack of objectivity of such psychological types. Thus a hazardous type overestimates the importance of the excess of buyers demand ($\Theta^i>0$) and underestimates the importance of the shortage of buyers demand ($\Theta^i<0$). A cautious type overestimates the importance of the shortage and underestimates the importance of the excess. The objective of each seller automaton is to find a price strategy which ensures on the average the least harmful situation (according to its psychology) created by the mis-match between commodity supply and demand in monetary units. Hence, each seller attempts to minimize the function (4) where the indicator function δ^i is given by eqn. (5).

The automata scheme (15) is considered to simulate the behavior of the sellers.

The buyers representing the seller's environment may be simulated by the "pair-wise comparison" model of section 6.1. In this case the utility of the j - th buyer making his purshase from the i - th seller is given by

$$\rho^j(y^i, y^k) = \alpha(y^k - y^i) \tag{54}$$

where α is some positive parameter.

The ψ function in eqn. (42) may be taken thus,

$$\psi(x) = \begin{cases} 1 & , \quad x > \Delta \\ (x+\Delta)/2\Delta, & -\Delta < x < \Delta \\ 0 & , \quad x < -\Delta \end{cases} \tag{55}$$

Here $(-\Delta, \Delta)$ represents the "active zone of the function". The function $\omega(.)$ in eqn.

(44) may be considered as,

$$\omega(y^i) = y^i \tag{56}$$

and ξ the amount of money available to each buyer.

3.7.1. Experiment 1 (Pairwise Comparison).

Several simulation experiments were carried out on a digital computer. The following numerical values are considered in the simulations,

Number of sellers $N = 3$, Sellers' psychology :
Number of buyers $\nu = 12$, Cautious $a^i = -1.$
$b^i = -0.005$
Active zone $\Delta = 10$, Objective $d^i = 0.02$
Hazardous $a^i = 1.$
$b^i = 0.005$

Commodity supply $q^1 = 2,\ q^2 = 2,\ q^3 = 3$
Available money to each buyer $\xi = 150$
Buyers' utility $\alpha = 0.05$, see eqn. (54). The sequence $\gamma^i(t)$, see eqn. (15), was taken as

$$\gamma^i(t) = \frac{\gamma_0}{t}\ ,\ \gamma_0 = \text{cont.},\ t = 1, 2, \ldots,\ (i = 1, 2, 3) \tag{57}$$

The automata sheme (15) always converged to certain equilibrium price probabilities independently of any initial assumptions.

For objective sellers and the following set of prices y_k^i

k \ i	1	2	3
1	100	100	140
2	140	130	170
3	-	170	200

The equilibrium (average) price probabilities were found as follows

$E\{p_k^i(100)\}$

k \ i	1	2	3
1	.0528	.0010	.6708
2	.9472	.4972	.3292
3	-	.5018	0

The probability of the first price for the third seller versus time is shown in Fig.3.

Fig.3 demonstrates the influence of γ_0 on the speed of convergence of the automata reinforcement scheme. It is concluded that a very small value of γ_0 (i.e. $0 < \gamma_0 << 1$) leads to a very sluggish convergence. On the other hand a value of γ_0 as big as 1 leads to a rather vigorous and oscillatory convergence. A suitable value for γ_0 was found to be somehow in between.

The effect on convergence rate that γ_0 has can be deduced from the form of the scheme (15). Note that the coefficients of $p^i_{k^i+u^i}(t)$, $p^i_j(t)$ are all unity. Any convergence will come from the "forcing terms" ; all of which are multiplied by $\gamma^i(t+1)$.

As in any expedient scheme, the parameter γ_0 contributes to determining the "degree of expediency", and consequently the value of the limiting probabilities, see e.g. Viswanathen and Narendra[14]. It is actually the ordering on the ensemble of strategies, rather than the precise magnitudes of the limiting probabilities which matters here. This ordering is insensitive to the choice of γ_0 provided that it is neither too big to induce premature convergence nor too small to impede the learning effect of the forcing terms. This is demonstrated in Experiment 2.

Compared to the objective case the hazardous sellers tend at equilibrium to increase the probability of higher prices.

The equilibrium probabilities are shown below :

$E\{p_k^i(100)\}$

k \ i	1	2	3
1	.0457	0	.3265
2	.9543	.1639	.6735
3	-	.8361	0

For the case of cautious sellers the equilibrium price probabilities show the tendency of increasing the probability of lower prices

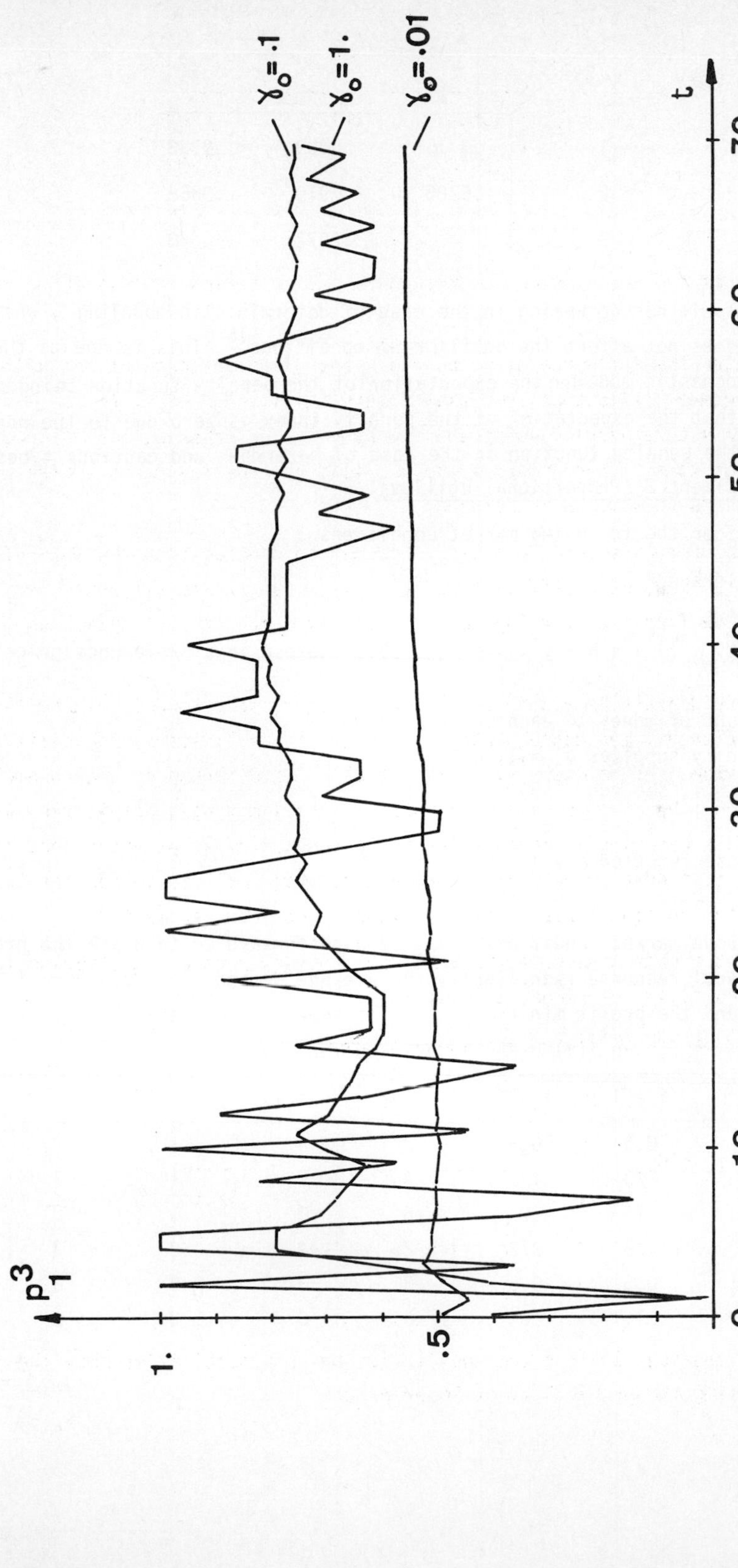

Fig.3 - Third-seller's first price probability versus time

$E\{p_k^i(100)\}$

k \ i	1	2	3
1	.3715	.0009	.9332
2	.6285	.7919	.0668
3	-	.2072	0

The above result has no analog in the case of deterministic modeling , where the psychology does not affect the equilibrium conditions[15]. This is due to the fact that in stochastic modeling the expectation of the penalty function being zero does not imply that the expectation of the penalty index is zero due to the non-linear form of the penalty function in the case of hazardous and cautious types.

3.7.2. Experiment 2 (Proportional utility).

Let us consider the following market conditions :

- number of sellers = 2
- number of buyers = 3
- buyer's utility $\rho^j(y^i) = h - y^i$, see eqn. (50) where h is a "reference price" for the buyer
- available amount of money to each buyer $\xi = 3$
- rate of commodity supplies $q^1 = 1$, $q^2 = 2$
- active zone $\Delta = 1$, see eqn. (55)
- set of prices, $Y^1 = \{1, 2\}$, $Y^2 = \{1, 2, 3\}$
- buyer's response function $\omega(y^i) = y^i$
- all sellers and buyers are objective.

Under the above market conditions it is straightforward to tabulate the probability of the buyers' response (eqn. (42)), the average demand $\pi^i = E\{s^i\} = \Sigma_{s^i} s^i p^i(s^i/\underline{y})$, and the profit min $(\pi^i, q^i y^i)$ as shown in the table below.

$\underline{y}$		$p^{ij}(y)$		aver. demand		Profit	
y^1	y^2	i = 1	i = 2	$\pi^1(y)$	$\pi^2(y)$	i = 1	i = 2
1	1	0,5	0,5	3	3	1	2
1	2	2/3	1/3	4	2	1	2
1	3	1	0	6	0	1	0
2	1	1/3	2/3	2	4	2	2
2	2	0,5	0,5	3	3	2	3
2	3	1	0	6	0	2	0

It follows from the table above that the optimal prices to be adopted by the sellers are $y^{1*} = 2$, $y^{2*} = 2$. At these prices the profit of each seller is maximum.

We simulated the stochastic automata model (12)** starting from the initial state of absolute randomeness, i.e.

$$\underline{p}^1 = (0{,}5,\ 0{,}5)^T, \quad \underline{p}^2 = (\tfrac{1}{3}, \tfrac{1}{3}, \tfrac{1}{3})^T$$

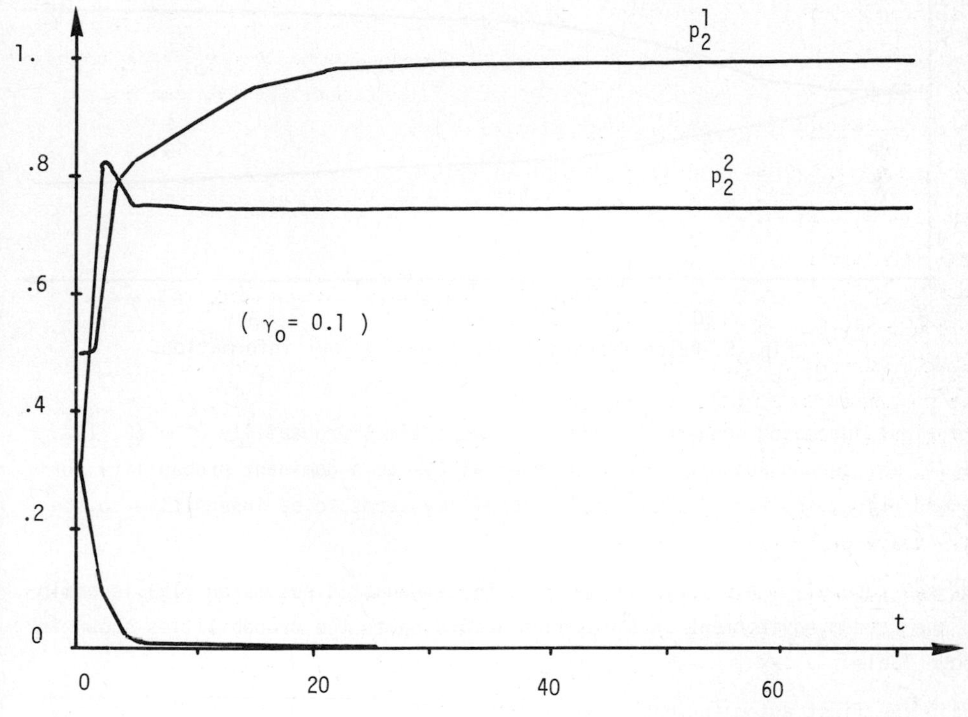

Fig. 4. Price probabilities - complete a priori information.

** Notice that the indicator function $\delta^i(\underline{y})$ is given by the difference between average demand π^i and the supply $q^i y^i$, i.e. $\delta^i(\underline{y}) = \pi^i(\underline{y}) - q^i y^i$

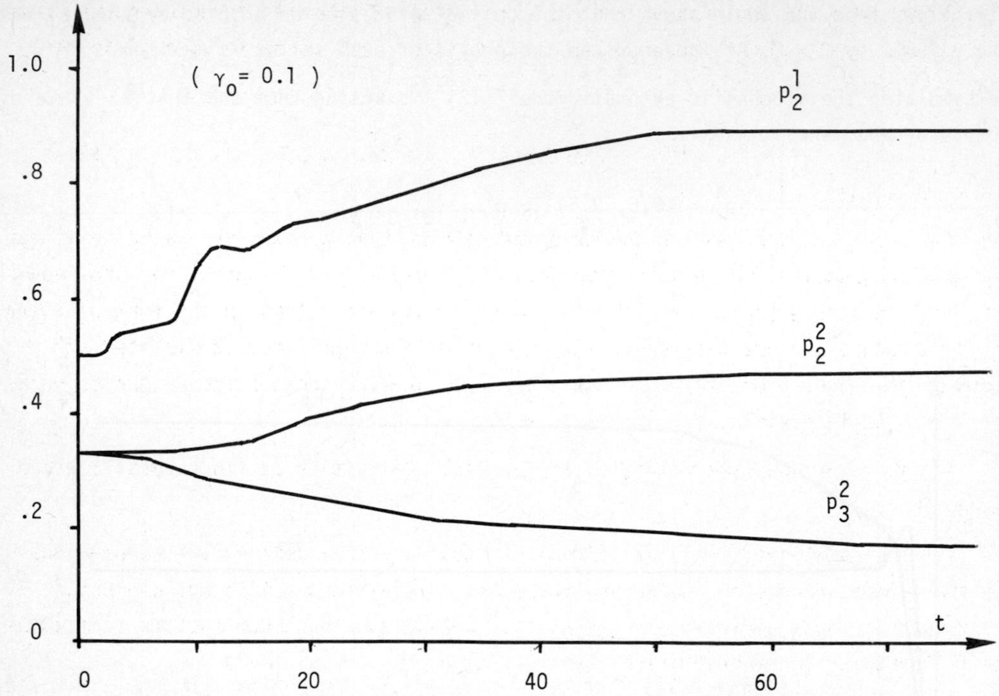

Fig. 5. Price Probabilities - No a priori information.

The first automaton converged rapidly to the optimal probability $\underline{\hat{p}}^1 = (0, 1)^T$, see Fig.4. The second automaton has converged always to a dominant probability for the second price, see Fig. 4. The final probability seems to be insensitive to the initial choice of γ^i.

For the case of no a priori information, the stochastic automaton (15) is employed in the random environment of the buyers according to the probabilities shown in the above Table.

Again the first automaton has converged to the optimal probability (0, 1), see Fig.5. The second automaton also converged to a probability that is dominant for the second price, with a value less than in the case of sufficient a priori information, see fig. 5. The ordering of the probability distribution is insensitive to the initial choice of γ^i.

3.8. RESOURCE ALLOCATION.

The problem of optimal allocation of a limited resource is formulated in the following way : there is a resource of quantity R with N users of the resource ; for each there is specified a function $\phi^i(s^i)$, this being the effect achieved by the i - th user of the resource if he uses quantity s^i of it. The effects achieved by the different users are commensurable, i.e., they are measured in homogeneous units. It is required to divide the existing resource among the users in such a way that the total effect is maximized, i.e.

$$\max_{s^1,s^2,\ldots,s^N} \sum_{i=1}^{N} \phi^i(s^i) \quad \text{subject to} \quad \sum_{i=1}^{N} s^i \leqslant R \text{ , } s^i \geqslant 0 \text{ } (i=1,..,N) \qquad (58)$$

The known computational procedures for solving that problem are based on either dynamic programming[16] or gradient methods [17]. Those methods assume prior knowledge of the functions $\phi^i(s^i)$. They lead to computational algorithms in the form of iterative procedures, where the constraint on available resource may be violated at intermediate computation steps. This makes direct on-line application of the computation results infeasible.

The above methods are not suitable for real application due to several reasons among them,

1 - The functions $\phi^i(s^i)$ are often not known a priori neither to the user nor to the allocation cneter. Moreover, the effects attained by different users can vary unpredictably during the relevant time period due to random factors like machine failure, varying market prices, etc.

2 - The users are active systems which use the information about their productiveness to promote their own goals[18].

It is of interest to explore the possibility of optimizing the system while it is in operation. That amounts to organizing collective behavior of the system, i.e. establishing the rules of interaction between the center and the users, formulating the criteria of the users, specifying their control variables,...etc.

We model the collective behavior of the N users as a game between N learning automata. The automata manipulate their control variable to promote their individual goals. Knowing only the realization of their "own" optimality criteria, the automata adapt control strategies as time unfolds. Successful adaptation must converge to an almost Nash-play of automata where it is to no user's advantage to change his control strategy. Organization of collective behavior is successful if the Nash play of automata corresponds to the desired optimality of the "over-all" resource-allocation system.

The collective behavior game is considered as follows. Initially the different users are given equal shares of the resource, i.e. $s^i(0) = R/N$. At the following planning periods (between the time steps $t = 1, 2,\ldots.$) each i - th unit receives a new amount of resource s^i with a change Δs^i from the amount obtained at the preceding period. Each i - th unit communicates to the center an estimate y^i of its production effectiveness at the current level of resource consumption s^i,

$$\phi^{i'}(s^i) \cong \frac{\Delta\phi^i}{\Delta s^i}\Big|_{s^i} \tag{59}$$

where $\Delta\phi^i$ is the change in production corresponding to an infitesimal change Δs^i of the resource.

The penalty index Θ^i of the i - th unit is considered to be the discrepancy between the actual and estimated production effectiveness,

$$\Theta^i = \phi^{i'}(s^i) - y^i \tag{60}$$

The penalty function of each i - th user is considered to be a positive function $|F^i(\Theta^i)|$. The penalty function $|F^i(.)|$ represents the law, set by the center to stimulate objective data from the units about their production effectiveness[18]. The following are two possible examples of that law,

$$F^i(\Theta^i) = \Theta^i|\Theta^i| \tag{61}$$

or

$$F^i(\Theta^i) = \begin{cases} \xi\Theta^i & , \quad \Theta^i < 0 \\ \eta\Theta^i & , \quad \Theta^i > 0 \end{cases} \qquad \xi, \eta > 0 \tag{62}$$

The nonlinear function $F^i(.)$ can also simulate psychological peculiarities of the producers,see sec. 7. Malishevskii[8] presented the organization of behavior in <u>continuous</u> form as follows,

<u>i - th user,</u>

$$\frac{d\,y^i}{d\,t} = \begin{cases} 0, \; (y^i{=}\phi^{i'}(0))\;(F^i{<}0)\;(y^i{=}\phi^{i'}(R))\;(F^i{>}0) \\ F^i(\phi^{i'}(s^i) - y^i)\;, \text{ otherwise} \end{cases} \tag{63}$$

<u>Center</u>

$$s^i = R\,\frac{g^i\,(y^i)}{\sum_{j=1}^{N} g^j(y^j)}, \; (i = 1, \ldots\ldots, N) \tag{64}$$

where the $g^j(y^j)$ are positive, continuously increasing functions of $y^j > 0$. $g^j(0) = 0$ for all j. The law (64) means that the environment allocates the resource on the elements in proportionality to their effectiveness. Malishevskii[8] showed that the system (63), (64) is stable and all trajectories $y^i(t)$ converge to the point,

$$\hat{y}^i = \frac{d\phi^i}{ds^i} \Big|_{\hat{s}^i(\hat{\underline{y}})}, \qquad (i = 1, \ldots, N) \qquad (65)$$

provided that ϕ^i are strictly concave functions for all $i = 1, \ldots, N$ and for all $0 \leqslant s^i \leqslant R$. The above organization yields a suboptimal solution of the resource allocation problem.

Optimality is approached when

$$y^1 \cong y^2 \cong \ldots\ldots \cong y^N \cong \text{const} = \hat{\lambda}$$

and the solution $\{s^i\}$, $\{y^i\}$ approaches the saddle point

$$\Phi(\hat{s}, \lambda) < \Phi(\hat{s}, \hat{\lambda}) < \Phi(s, \hat{\lambda})$$

where,

$$\Phi(s, \lambda) = \sum_{i=1}^{N} \Phi^i(s^i) + \lambda\left(R - \sum_{j=1}^{N} s^j\right)$$

Here we consider the stochastic automata analog of Malishevskii's model. The variables ϕ^i, $\phi^{i'}$ for any s^i are considered to be stochastic variables with unknown distributions. The estimation y^i of the production effectiveness $\phi^{i'}$ is considered to be the automaton's action or strategy. The learning of each i - th producer-automaton is directed towards decreasing its own penalty,

$$|\delta^i(\underline{y})| = E\{|F^i(\phi^{i'}(s^i) - y^i)|\} \qquad (66)$$

where s^i is given by (64). Notice that the above formulation corresponds to a Nash game : any realization of δ^i satisfies the contramonotonicity property.

To show this, let $y^{i(1)} > y^{i(2)}$, then

$$s^{i(1)} = \frac{g^i(y^{i(1)})}{g^i(y^{i(1)}) + \sum_{j \neq i} g^j(y^j)} > \frac{g^i(y^{i(2)})}{g^i(y^{i(2)}) + \sum_{j \neq i} g^j(y^j)} = s^{i(2)}$$

According to the natural assumption concerning the diminution of production effectiveness with the increase of the used quantity of resources, we have $\phi^{i'}(s^{i(1)}) < \phi^{i'}(s^{i(2)})$ whenever $s^{i(1)} > s^{i(2)}$. Hence

$$\phi^{i'}\left(\frac{g^i(y^{i(1)})}{g^i(y^{i(1)})+\sum_{j\neq i} g^j(y^j)}\right) - y^{i(1)} < \phi^{i'}\left(\frac{g^i(y^{i(2)})}{g^i(y^{i(2)})+\sum_{j\neq i} g^j(y^j)}\right) - y^{i(2)}$$

and consequently

$$F^i(\phi^{i'}(.) - y^{i(1)}) < F^i(\phi^{i'}(.) - y^{i(2)})$$

Computer simulation were carried out for ten consumers with the following production functions,

$$\phi^1(s^1) = 4,25\ \ell n\ (1+s^1) \qquad \phi^6(s^6) = 2 \sin s^6$$
$$\phi^2(s^2) = 2,125\ \ell n\ (1+s^2) \qquad \phi^7(s^7) = s^7\ (2,125-s^7)$$
$$\phi^3(s^3) = \sqrt{2}\ s^3 \qquad \phi^8(s^8) = \frac{1}{2}\ s^8(4,25-s^8)$$
$$\phi^4(s^4) = \sqrt{3}\ s^4 \qquad \phi^9(s^9) = 2,25\ (1-e^{-s9})$$
$$\phi^5(s^5) = \frac{6}{\pi} \sin \frac{\pi}{3}\ s^5 \qquad \phi^{10}(s^{10}) = 2,25\ (1-e^{-2s10})$$

The set of trategies for each automaton taken as the values of $\phi^{i'}$ at ten points between 0 and 1 ; R is taken to equal one. The functions $F^i(.)$ were taken as in (62) with $\xi = \eta = 1$. The functions $g^i(y^i)$, see (64), were taken as

$$g^i(y^i) = (y^i)^r, \qquad (i = 1, \ldots\ldots, N) \qquad (67)$$

The sequences $\gamma(t)$, $\varepsilon(t)$, see (28) were taken as γ_0/t and ε_0/t, respectively.

Convergence was always observed after short time interval. The algorithm (28) demonstrated low sensitivity to the choice of the parameters γ_0, and ε_0. Fig. 6 depicts the total production versus time. The results are improved when r gets larger, see Fig. 6. This means that if the rule of distribution of the resource is close to the rule "provide to him who gives the maximum estimate of effectiveness" the solution approches the optimal one[8]. The organization however in the latter case seems to be more sensitive to elements failure, see Fig. 6.

At time t^*, see Fig. 6 the first producer was considered to break down and start to emit the estimate $y^1 = 0$. That instantly caused the total production to drop off drastically. Immediately then the system self-organized itself in such a way that the resource was redistributed among the remaining producers so that the increase in effects of the remaining producers partially compensated the drop in total production. This result is interesting as it demonstrates the reliability of the system.

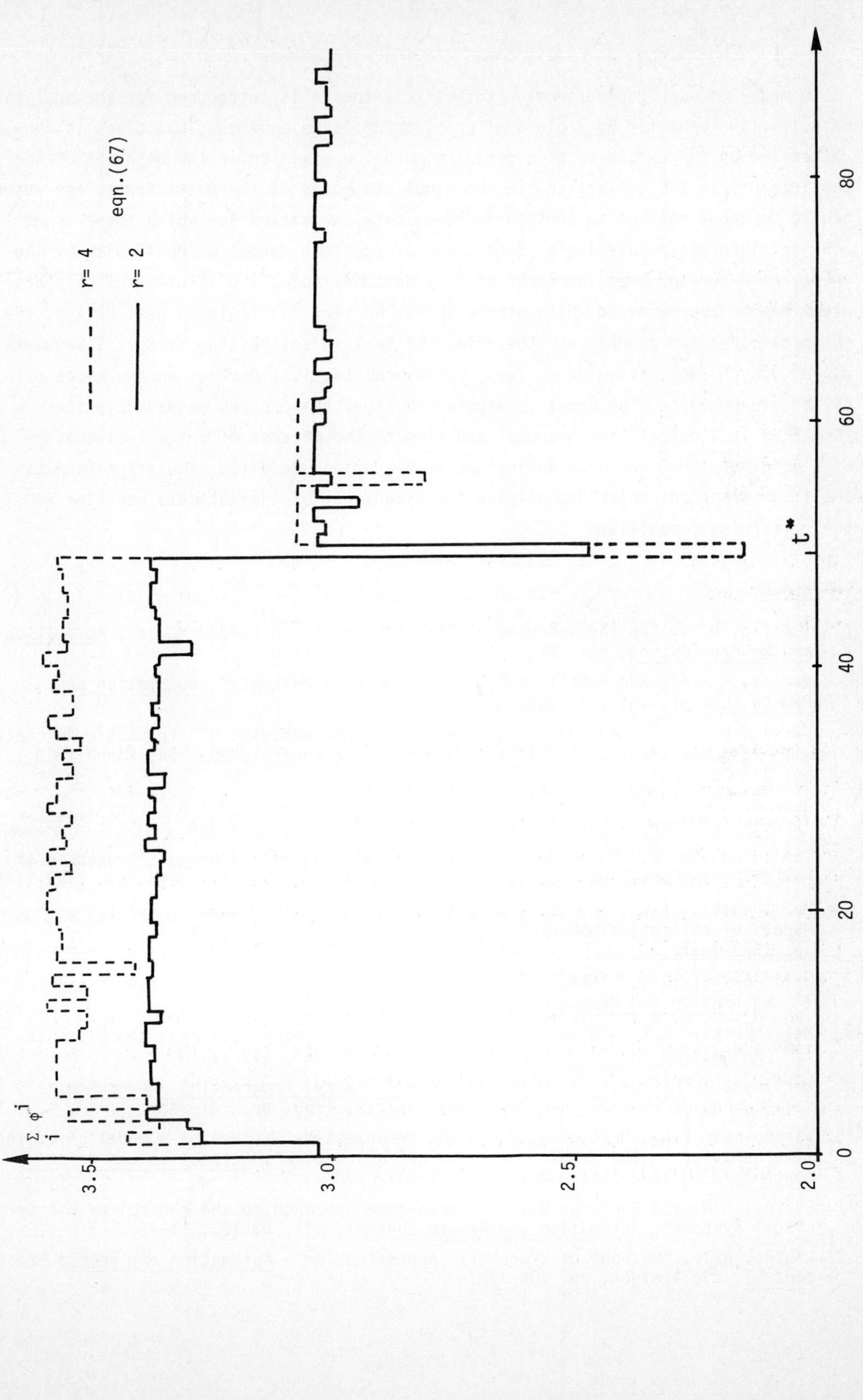

Fig. 6. Production versus time.

3.9. CONCLUSIONS.

A model of many goal-oriented stochastic automata is introduced for the analysis of collective behavior of a class of problems of large systems. That class is characterized by the existence of a definite mutual opposition in the behavior of the participants in the collective. In the model the goals of the participants are assumed to be known only up to certain indeterminate parameters for which there is no a priori information available. Such class of problems cannot be dealt with by the theory of N-person games. By means of that automata model the solution of such problems may be approximated which otherwise may be very difficult to get. Besides the automata model can demonstrate the effect of certain interesting factors like participants psychology, stimulation laws, behavioral tactics, etc.... on the modes of collective behavior. The model is applied to study the process of market price formation in a competitive economy, and also to the process of optimal allocation of a unidimensional resource during system operation. Detailed simulation results are also presented, which demonstrate the expediency of the automata behavior and their learning capability.

REFERENCES.

1. Tsetlin, M.L., "On the behavior of Finite Automata in Random Media", Automation and Remote Control, vol. 22, Oct. 1961, pp. 1210-1219.
2. Krylov, V. Yu., and Tsetlin M.L., "Games between automata", Automation and Remote control, vol. 24, July 1962, pp. 889-899.
3. Varshavskii, V.I., and Vorontsova, I.P., "On the behavior of Stochastic Automata with Variable Structure", Automation and Remote Control, vol. 24, March 1963, pp. 327-333.
4. Narendra, K.S., and Thathachar, M.A.L., "Learning Automata : a survey", IEEE Trans. Syst. Man, Cybern., vol. SMC-4, N°4, 1974, pp. 323-334.
5. Tenisberg, Yu. D., "Some Models of collective behavior in Dynamic Processes of Market Price Formation", Automation and Remote Control, n°7, 1969, pp. 1140-1148.
6. Malishevskii, A.V., and Tenisberg, Yu.D., "One class of Games connected with Models of collective behavior", Automation and Remote Control, n°11, 1969, pp. 1828-1837.
7. Malishevskii, A.V., "Models of Joint Operation of many Goal-Oriented Elements, 1", Automation and Remote Control, n°11, 1925-1845, (1971).
8. Malishevskii, A.V., "Models of joint operation of many Goal-oriented Elements, II", Automation and Remote Control, n°12, 2020-2028, (1971).
9. Karlin, S., "Mathematical Methods in Games Theory, Programming and Economics", vol. 1, Addison-Wesley, Reading Massachusetts, 1959, pp. 301-348.
10. Varshavskii V.I., Meleshina, M.V., and Perekrest V.T., "Use of a model of collective behavior in the problem of resources allocation", Automation and Remote Control, n°7, 1107-1114, 1969.
11. Burkov, V.N. and Opoitsev V.I., "A meta-game approach to the control of Hierarchical systems", Automation and Remote Control, n°1, 93-103, (1973).
12. Loginov N.V., "Methods of stochastic approximation", Automation and Remote control, 27, 4, 1966, pp. 706-728.

13. Robbins H., and Siegmund D., "A convergence theorem for nonnegative almost supermartingales and some applications", in Optimizing Methods in Statistics", pp. 233-257. Academic Press, New York, 1971.

14. Viswanathan R., and Narendra K.S., "Comparison of Expedient and Optimal Reinforcement Schemes for Learning Systems", Journal of Cybernetics, vol. 2, n°1, 1972, pp. 21-37.

15. Krylatykh L.P., "On a Model of Collective Behavior", Engineering Cybernetics, 1972, pp. 803-808.

16. Bellman R., and Dreyfus S., Applied Dynamic Programming, Princeton : University Press, 1962.

17. Arrow K., Hurwitz L. and Uzawa H., Studies in linear and nonlinear Programming, Standford : California Stanford University Press, 1958.

18. Ivanovskii A.G., "Problems of stimulation and obtaining objective Estimates in active systems", Automation and Remote Control, N°8, 1298-1303, 1970.

19. El Fattah Y.M., "Learning Automata as models of behavior", Simulation 75 Proceedings, Zurich (Switzerland) (1975).

20. El Fattah Y.M., "A model of many goal-oriented stochastic automata with application to a marketing problem", 7th IFIP Conf. on Optimization Techniques proceedings, Nice (France) (1975).

21. El Fattah Y.M., "Analysis of collective behavior in large systems using a model of many goal-oriented stochastic automata with applications", IFAC Symp. on large-scale systems proceedings, Udine (Italy) (1976).

22. El Fattah Y.M., and R. Henriksen, "Simulation of market price formation as a game between stochastic automata", J. Of Dynamic Systems, Measurement and control (special issue) March (1976).

23. El Fattah Y.M., "Use of a learning automata model in resource allocation problems", IFAC Symp., Cairo (Egypt) 1977.

A P P E N D I X

Projection operator

The N-dimensional simplex $S_\varepsilon = \{\underline{y} : \sum_{i=1}^{N} y^i = 1,\ y_i > \varepsilon,\ \underline{y}\varepsilon R^N\}$ can be transformed into the simplex $S = \{\underline{x} : \sum_{i=1}^{N} x_i = L,\ x_i > 0,\ \underline{x}\varepsilon R^N\}$ by means of the simple change of variables

$$y_i = x_i + \varepsilon \quad , \; i = 1,\ldots, N \tag{A.1}$$

We stipulate that

$$\varepsilon < \frac{1}{N} \tag{A.2}$$

in order that $L = 1 - N\varepsilon$ be a positive number. The N-dimensional simplex has the following for $N \geqslant 3$:

N vertices $\{T_1^j\}$

N(N-1)/2 edges (two-dimensional faces) $\{T_2^k\}$

....

N (N-1)dimensional faces $\{T^{\ell}_{N-1}\}$

The vertex T^j_1 is the point

$$T^j_1 = (\underbrace{0, 0, \ldots, L}_{j}, 0, \ldots, 0) \tag{A.3}$$

and the face T^k_m $(m \geq 2)$ is the subset

$$T^k_m = \{\underline{x} : \underline{x} \varepsilon D_m, x_i \geqslant 0 \forall i\} \tag{A.4}$$

of one of the hyperplanes D_m :

$$\sum_{i=1}^{N} a_i x_i = L \tag{A.5}$$

where $a_i \varepsilon \{0, 1\}$ and

$$\sum_{i=1}^{N} a_i = m \tag{A.6}$$

Obviously, for a particular $m(1<m\leqslant N)$ there are $\frac{N!}{m!(N-m)!}$ realizations of such hyperplanes.

The projection of $\underline{x} \varepsilon R^N$ into the plane $D_m(1<m\leqslant N)$ is accomplished according to the formula

$$(\underline{x}(D_m))_j = (x_j + \frac{L - \sum_{i=1}^{N} a_i x_i}{m}) a_j \qquad (j = 1, \ldots, N) \tag{A.7}$$

In order to see that $\underline{x}(D_m)$ lies on D_m, perform the summation of both sides of (A.7) after premultiplying into a_j to get

$$\sum_{j=1}^{N} a_j(\underline{x}(D_m))_j = \sum_{j=1}^{N} a_j^2 x_j + \sum_{j=1}^{N} a_j^2 (\frac{L - \sum_{i=1}^{N} a_i x_i}{m}) = \sum_{j=1}^{N} a_j x_j + m(\frac{L - \sum_{i=1}^{N} a_i x_i}{m})$$

$$= L$$

Here equation (A.6) is employed.

Let us now state the following lemma.

<u>Lemma</u>. The face T_{m-1} closest to the point $\underline{x}(D_m)$ has an orthogonal vector $\underline{a}_{m-1}$ with the components

$$(a_{m-1})_k = \begin{cases} (a_m)_k & k \neq j \\ 0 & k = j \end{cases} \tag{A.8}$$

where the index j corresponds to the minimal component of the point $x(D_m)$, that is

$$(\underline{x}(D_m))_j = \min_i (\underline{x}(D_m))_i \tag{A.9}$$

Proof. The distance V from a certain point $\underline{z}\varepsilon D_m$ to the vertex T_1^k is equal to

$$V^2(\underline{z}, T_1^k) = (z_k - L)^2 + \sum_{\substack{i=1 \\ i\neq k}}^{N} z_i^2 = \sum_{i=1}^{N} z_i^2 + (L^2 - 2z_k L)$$

Then

$$V^2(\underline{z}, T_1^k) - V^2(\underline{z}, T_1^m) = 2(z_m - z_k)L \tag{A.10}$$

i.e. the most distant vertex corresponds to the least component $z_k = \min_m z_m$. Consequently the face T_{m-1} which lies opposite to the vertex and closest to the point $\underline{z}$ has the orthogonal vector $\underline{a}_{m-1}$ obtained by nullifying the j-th component of $\underline{a}_m$, see (A.8). The lemma has been proved.

By definition, the projection $\Omega(\underline{x}^\circ)$ of a point $\underline{x}^\circ \varepsilon R^N$ into S is called the point

$$\Omega(\underline{x}^\circ) = \{\underline{x}^* : ||\underline{x}^\circ - \underline{x}^*|| = \min_{\underline{y}\varepsilon S} ||\underline{x}^\circ - \underline{y}||\} \tag{A.11}$$

It follows then that finding $\underline{x}^* = \Omega(\underline{x}^\circ)$ is equivalent to finding the point on S which is closest to the projection $\underline{x}^\circ(D_m)$ of the point $\underline{x}^\circ$ into $D_m (1<m<N)$, provided that $\underline{x}^\circ(D_k) \not\varepsilon S \; \forall \; k \geq m$. Actually,

$$\min_{\underline{y}\varepsilon S} ||\underline{y} - \underline{x}^\circ||^2 = \min_{\underline{y}\varepsilon S} ||(\underline{y} - \underline{x}^\circ(D_m)) + (\underline{x}^\circ(D_m) - \underline{x}^\circ)||^2$$

$$= ||\underline{x}^\circ(D_m) - \underline{x}^\circ||^2 + \min_{\underline{y}\varepsilon S} ||\underline{y} - \underline{x}^\circ(D_m)||^2 \tag{A.12}$$ (1)

The property (A.11) of the projection operator as well as the Lemma suggest the following sequential procedure for determining the projection,

a) check the condition $\underline{x}^\circ \varepsilon S$.

b) if $x^\circ \not\varepsilon S$ then find the projection $\underline{x}^\circ(D_N)$

c) if $\underline{x}^\circ(D_N) \not\varepsilon S$ then find the face T_{N-1}^k closest to the point $\underline{x}^\circ(D_N)$ from the Lemma,

---

(1) Notice that $\langle \underline{x}^\circ - \underline{y}, \underline{y} \rangle = 0$ if $\underline{y}$ is the projection of $\underline{x}^\circ$. Also $\langle \underline{x}^\circ(D_m) - \underline{x}^\circ, \underline{x}^\circ(D_m) \rangle = 0$. Hence if $\underline{y}$ is the projection of $\underline{x}^\circ(D_m)$ then $\langle \underline{x}^\circ(D_m) - \underline{y}, \underline{y} \rangle = 0$ and consequently $\langle \underline{x}^\circ(D_m) - \underline{x}^\circ, \underline{y} - \underline{x}^\circ(D_m) \rangle = 0$

d) project $\underline{x}^\circ(D_N)$ into $D_{N-1} \supset T^k_{N-1}$. If $\underline{x}^\circ(D_{N-1}) \notin S$ then we find the face T^j_{N-2} closest to the point $\underline{x}^\circ(D_{N-1})$ from the Lemma and again project $\underline{x}^\circ(D_{N-1})$ into $D_{N-2} \supset T^\ell_{N-2}$ and so forth till $\underline{x}^\circ(D_m) \varepsilon S$ for certain m.

e) if $\underline{x}^\circ(D_2) \notin S$ then the projection will be one of the vertices T^r_1.

According to the above procedure the projection operator Ω is executed at N steps at most.

CHAPTER IV

CONTROL - FINITE MARKOV CHAINS.

"Un jour tout sera bien, voilà notre espérance,
tout est bien aujourd'hui, voilà l'illusion".
Voltaire. Poème sur le désastre de Lisbonne

4.1 MARKOV DECISION MODEL.

A Markov decision model is first described. A system is observed at equally spaced epochs numbered 0, 1, 2, ... At each epoch n the system is observed to occupy one of N states numbered 1 through N. Each state i has associated with it a finite set K_i of M_i decisions. Whenever state i is observed some decision k in K_i must be selected. Suppose state i is observed at epoch n and decision k is selected. The probability that state j is observed at epoch n+1 is $\pi(i, k, j)$. Transitions are presumed to occur with probability 1, so that

$$0 \leq \pi(i,k,j) \quad \text{and} \quad \sum_{j=1}^{N} \pi(i,k,j) = 1 \ , \ i=1,\dots, N, \ K \varepsilon K_i \qquad (1)$$

We assume a certain reward structure superimposed on the Markovian decision process. Whenever the system is in state i and decision $k \varepsilon K_i$ is taken, then a reward r_{ikj} will be earned upon transiting to a j-th state.

A control policy D is a prescription for making decisions at each epoch. We shall consider only the class of stationary memoryless policies. A policy in such class is specified by a set of N vectors $\underline{d}^{(1)}, \underline{d}^{(2)}, \dots, \underline{d}^{(N)}$. The vector $\underline{d}^{(i)}$ is M_i dimensional. Its k-th component $d_k^{(i)}$ specifies the probability for making the k-th decision in K_i whenever observing the i-th state. Naturally, $\underline{d}^{(i)}$ must belong to the simplex

$$S^{M_i} = \{\underline{x} \varepsilon R^{M_i} : x_j \geq 0, \ \sum_{j=1}^{M_i} x_j = 1\} \quad , \qquad i = 1, \dots N \quad (2)$$

The subclass of policies for which $d_k^{(i)}$ is zero except for exactly one $k \varepsilon K_i$, for every i, is the class of deterministic policies.

We note that for any fixed D the observed states $\{x_n\}_{n \geq 0}$ constitute a homogeneous Markov chain whose transition matrix $P(D) = (P_{ij}(D))$ is given by

$$P_{ij}(D) = \sum_{k\in K_i} \pi(i,k,j)\, d_k^{(i)}, \qquad i,j = 1, \ldots, N \tag{3}$$

For any fixed D the chain is assumed to be ergodic. That means that the chain is characterized by one irreducible closed set of persistent aperiodic states. Transient states are allowed and can vary with the policy.

Definition. Let P be a stochastic matrix. The ergodic coefficient of P, denoted by $\alpha(P)$, is defined by

$$\alpha(P) = 1 - \sup_{i,k} \sum_{j=1}^{N} (P_{ij} - P_{kj})^{+} \tag{4}$$

where $(P_{ij} - P_{kj})^{+} = \max(0, P_{ij} - P_{kj})$.

A homogeneous Markov chain is ergodic if and only if $\alpha(P^k) > 0$ for some k, cf. Isaacson and Madsen[1].

Under the ergodicity assumption there exists exactly one long-run state distribution $P_i(D)$, $i = 1, \ldots, N$, satisfying the conditions,

i. $p_i(D) \geq 0$

ii. $p_j(D) = \sum_{i=1}^{N} p_i(D) P_{ij}(D)$

iii. $\sum_{j=1}^{N} p_j(D) = 1$

Let $\underline{p}^T(D)$ be the 1 x N vector $(p_1(D), \ldots, p_N(D))$ and $\underline{1}$ be the N x 1 vector of 1's. Then $\underline{p}^T(D)$ is the solution of

$$\underline{p}^T(D)(I - P(D)) = \underline{0}, \quad \underline{p}^T(D)\underline{1} = 1 \tag{5}$$

which is a system of N+1 equations in N unknowns. Since it specifies $\underline{p}(D)$ uniquely, it must contain exactly one redundancy. Since P is stochastic (by virtue of eqs.(1) and (2)) the columns of (I-P(D)) sum to zero, i.e.

$$(I - P(D))\,\underline{1} = \underline{0} \tag{6}$$

So any of the columns of (I - P(D)) can be eliminated. Let B(D) be the NxN matrix obtained by replacing the first column of (I - P(D)) by $\underline{1}$. Then eqs. (5) and (6) combine to assure that $\underline{p}(D)$ is the unique solution of

$$\underline{p}^T(D)\, B(D) = \underline{e}_1^T \tag{7}$$

where $\underline{e}_1^T$ is the 1 x N vector (1, 0, ..., 0). B is invertible due to the fact that it is of rank N. Let Q(D) denote the inverse of B(D). It follows from (7) that

$$\underline{p}^T(D) = \underline{e}_1^T\, Q(D) \tag{8}$$

That is $\underline{p}^T(D)$ is the top row of Q.

The assumption that the chain is ergodic for all D implies that the expected reward does not depend on the initial state i after a sufficiently long interval of time. The long-run expected average reward can be expressed as

$$\phi(D) = \sum_{i=1}^{N} \sum_{k \varepsilon K_i} d_k^{(i)} \eta_{ik} \, p_i(D) \tag{9}$$

where,

$$\eta_{ik} = \sum_{j=1}^{N} \pi(i,k,j) \, r_{ikj} \quad , \quad i=1,\ldots,N \; ; \; k \varepsilon K_i \tag{10}$$

is the expected average reward per stage when the state i is observed and the k-th decision in K_i is taken. Without loss of generality we assume that all the η's are nonnegative. They must be subject to a finite upper bound,

$$0 \leq \eta_{ik} \leq c_1 < \infty \qquad \text{all } i,k \tag{11}$$

The control problem is to find the optimal policy D which maximizes the expected average reward $\phi(D)$ subject to the system equation (7).

4.2 CONDITIONS OF OPTIMALITY.

A variational approach is adopted to find the conditions of optimality. An admissible variation $\delta_{\alpha\beta}(\Delta)$ of a policy D is defined by the system of N vectors,

$$\delta^i_{\alpha\beta}(\Delta) = \begin{cases} \Delta \, \underline{e}^{(\alpha)}_{\beta} & , \; i = \alpha \\ \underline{0}_{M_i} & , \; \text{otherwise} \end{cases} \qquad \alpha = 1,\ldots, N \; ; \; \beta = 1,\ldots, M_\alpha \tag{12}$$

where,

$$\underline{e}^{(\alpha)T}_{\beta} = (\underbrace{-\frac{1}{M_\alpha - 1}, \ldots, -\frac{1}{M_\alpha - 1}}_{\beta - 1}, 1, -\frac{1}{M_\alpha - 1}, \ldots, -\frac{1}{M_\alpha - 1}) \qquad \begin{matrix} \alpha=1,\ldots N \\ \beta=1,\ldots M_\alpha \end{matrix} \tag{13}$$

and $\underline{0}_{M_i}$ denotes the M_i dimensional null vector. Here $\underline{\delta}^i_{\alpha\beta}(\Delta)$ denotes the perturbation of the policy vector $\underline{d}^{(i)}$.

The variation "step-lenght" Δ, eqn. (12), must satisfy the condition that $\underline{d}^{(\alpha)} + \delta^{\alpha}_{\alpha\beta}(\Delta)$ lie in the simplex $S^{M}\alpha$. This amounts to

$$\begin{aligned} &\Delta < \min\,(1 - d^{(\alpha)}_{\beta}, (M_\alpha - 1) \min_{k \neq \beta} d^{(\alpha)}_k), \quad \Delta > 0 \\ &|\Delta| < \min\,(d^{(\alpha)}_{\beta}, (M_\alpha - 1)(1 - \max_{k \neq \beta} d^{(\alpha)}_k)), \quad \Delta < 0 \end{aligned} \tag{14}$$

An admissible variation $\delta_{\alpha\beta}(\Delta)$ of a policy D leads to a corresponding variation $\delta_{\alpha\beta}B$ of the matrix B(D) of eqn. (7). That variation is given by,

$$(\delta_{\alpha\beta}B)_{ij} = \begin{cases} 0 \ , \ i \neq \alpha \ , \ \text{all } j \\ \Theta^{j}_{\alpha\beta} \ \Delta, i=\alpha, \ j=1,\ldots, N \end{cases} \tag{15}$$

where,

$$\Theta^{j}_{\alpha\beta} = \begin{cases} 0, \ j=1 \\ -\pi(\alpha,\beta,j) + \sum_{k\neq\beta} \frac{1}{M_\alpha - 1} \pi(\alpha,k,j), \ j=2,\ldots, M \end{cases} \tag{16}$$

Using a matrix inversion lemma, cf. Durand[2], we can write the inverse of the matrix $B + \delta_{\alpha\beta}B$ as

$$(B + \delta_{\alpha\beta}B)^{-1} = Q + \delta_{\alpha\beta}Q \tag{17}$$

where Q is the inverse of B and $\delta_{\alpha\beta}Q$ is the matrix

$$(\delta_{\alpha\beta}Q)_{ij} = - \frac{Q_{i\alpha} \ \xi^{j}_{\alpha\beta}}{1+ \Delta\xi^{\alpha}_{\alpha\beta}} \ , \quad i,j=1,\ldots,N \tag{18}$$

and

$$\xi^{j}_{\alpha\beta} = \sum_{i=1}^{N} \Theta^{i}_{\alpha\beta} Q_{ij} \ , \quad j=1,\ldots,N \tag{19}$$

Let us adopt the following definitions of the norm of a vector $\underline{a}$ and a matrix A,

$$||\underline{a}|| = \max_{i} |a_i| \qquad ||A|| = \max_{i,j} |A_{ij}| \tag{20}$$

Since Q is the inverse matrix of B, it follows from the Schwarz inequality that

$$1 = ||B.Q|| \leq ||B||.||Q|| = ||Q|| \leq c_o < \infty \tag{21}$$

It also follows from eqn. (19) that

$$||\xi_{\alpha\beta}|| \leq ||Q|| \ ||\underline{\Theta}_{\alpha\beta}|| \leq c_o \max_{j} \ (\max_{k} \pi(\alpha,k,j) - \min_{k}\pi(\alpha,k,j)) = c(\alpha) \leq c_o \tag{22}$$

The constants c_o and $c(\alpha)$ can always be chosen big enough to have inequality (22) valid for all policies D.

Eqs. (8), (18) and (19) combine to assure that the variations $\delta_{\alpha\beta}p_i(D)$ and $\delta_{\alpha\beta}\xi^{i}_{\alpha\beta}(D)$,

$i=1,\dots,N$ are given by,

$$\delta_{\alpha\beta}p_i = -\Delta . p_\alpha \frac{\xi^i_{\alpha\beta}}{1 + \Delta . \xi^\alpha_{\alpha\beta}} \;, \quad i=1,\dots, N \tag{23}$$

$$\delta_{\alpha\beta}\ \xi_{\alpha\beta} = -\Delta . \xi^\alpha_{\alpha\beta} \frac{\xi^i_{\alpha\beta}}{1 + \Delta . \xi^\alpha_{\alpha\beta}} \;, \quad i=1,\dots, N \tag{24}$$

Hence the variation in the expected reward, eqn. (9), resulting from a variation $\delta_{\alpha\beta}(\Delta)$ of a policy D, eqn. (12), can be written as

$$\begin{aligned}\delta_{\alpha\beta}\phi(D) &= \sum_{i=1}^{N} \sum_{k=1}^{M_i} (\delta_{\alpha\beta}d_k^{(i)} . \eta_{ik} . p_i + d_k^{(i)} . \eta_{ik} . \delta_{\alpha\beta}p_i) \\ &= \Delta(\eta_{\alpha\beta} - \frac{1}{M_\alpha - 1} \sum_{k\neq\beta} \eta_{\alpha k} - \sum_{i=1}^{N} \sum_{k=1}^{M_i} d_k^{(i)}\ \eta_{ik} \frac{\xi^i_{\alpha\beta}}{1+\Delta . \xi^\alpha_{\alpha\beta}}) p_\alpha\end{aligned} \tag{25}$$

We define the derivatives,

$$\begin{aligned}\frac{\partial\ \phi}{\partial\ d_\beta^{(\alpha)}}(D) &= \lim_{\Delta\to 0} \delta_{\alpha\beta}\phi(D)/\Delta \\ &= (\eta_{\alpha\beta} - \frac{1}{M_\alpha - 1} \sum_{k\neq\beta} \eta_{\alpha k} - \sum_{i=1}^{N} \sum_{k=1}^{M_i} d_k^{(i)}\ \eta_{ik}\ \xi^i_{\alpha\beta}) p_\alpha\end{aligned} \tag{26}$$

$\alpha=1,\dots, N$; $\beta \in K_\alpha$. The variation $\delta_{\alpha\beta}\phi$, eqn.(25), can be expanded as a function of Δ. Keeping only those terms with powers of Δ less than or equal to 2, we have

$$\delta_{\alpha\beta}\phi(D) = \frac{\partial\ \phi}{\partial\ d_\beta^{(\alpha)}}(D)\ .\ \Delta + \Psi_{\alpha\beta}(D)\ .\ \Delta^2 + o(\Delta^3) \tag{27}$$

where,

$$\Psi_{\alpha\beta}(D) = \sum_{i=1}^{N} \sum_{k=1}^{M_i} d_k^{(i)}\ \eta_{ik}\ \xi^i_{\alpha\beta}\ \xi^\alpha_{\alpha\beta}\ p_\alpha \tag{28}$$

Let us now examine the effect of a variation $\delta_{\gamma\delta}$ of a policy D on the first order derivatives $\partial\phi/\partial d_\beta^{(\alpha)}$ given by eqn.(26). That variation can be written as,

$$\begin{aligned}\delta_{\gamma\delta} \frac{\partial\ \phi}{\partial d_\beta^{(\alpha)}} &= (\eta_{\alpha\beta} - \frac{1}{M_\alpha - 1} \sum_{k\neq\beta} \eta_{\alpha k} - \sum_{i=1}^{N} \sum_{k=1}^{M_i} d_k^{(i)}\ \eta_{ik}\ \xi^i_{\alpha\beta})\ \delta_{\gamma\delta}p_\alpha \\ &\quad - \sum_{i=1}^{N} \sum_{k=1}^{M_i} d_k^{(i)} \eta_{ik}\ \delta_{\gamma\delta}\xi^i_{\alpha\beta}\ p_\alpha - \Delta(\eta_{\gamma\delta} - \frac{1}{M_\gamma - 1} \sum_{k\neq\delta} \eta_{\gamma k}) \xi^i_{\alpha\beta}\ p_\alpha\end{aligned} \tag{29}$$

Considering the variations $\delta_{\gamma\delta}$ of both sides of eqs.(8), (19), and making use of eqn.(18), we get

$$\delta_{\gamma\delta}\ p_{\alpha} = -\ \Delta.p_{\gamma}.\ \frac{\xi^{\alpha}_{\gamma\delta}}{1 + \Delta\ \xi^{\gamma}_{\gamma\delta}} \tag{30}$$

$$\delta_{\gamma\delta}\ \xi^{i}_{\alpha\beta} = -\ \Delta\ \xi^{\gamma}_{\alpha\beta}\ \frac{\xi^{i}_{\gamma\delta}}{1 + \Delta\ \xi^{\gamma}_{\gamma\delta}} \tag{31}$$

Substituting from (30) and (31) into (29) and then taking the limit of the division by Δ as $\Delta\to 0$, we get what we call the second-order derivatives,

$$\frac{\partial^2\ \phi}{\partial d^{(\gamma)}_{\delta}\partial d^{(\alpha)}_{\beta}} = \lim_{\Delta\to 0}\ \frac{1}{\Delta}\ \delta_{\gamma\delta}\ \frac{\partial\ \phi}{\partial d^{(\alpha)}_{\beta}} = -(\eta_{\alpha\beta} - \frac{1}{M_{\alpha}-1}\ \sum_{k\neq\beta}\eta_{\alpha k} - \sum_{i=1}^{N}\sum_{k=1}^{M_i} d^{(i)}_{k}\eta_{ik}\xi^{i}_{\alpha\beta})p_{\gamma}\xi^{\alpha}_{\gamma\delta}$$

$$-\ (\eta_{\gamma\delta} - \frac{1}{M_{\gamma}-1}\ \sum_{k\neq\delta}\eta_{\gamma k} - \sum_{i=1}^{N}\sum_{k=1}^{M_i} d^{(i)}_{k}\eta_{ik}\xi^{i}_{\gamma\delta})p_{\alpha}\xi^{\gamma}_{\alpha\beta}$$

$$\alpha,\gamma = 1,\ldots,N\ ;\ \beta\varepsilon K_{\alpha},\ \delta\varepsilon K_{\gamma} \tag{32}$$

It can be easily verified that any two arbitrary policies $\underline{d}^{(\alpha)}$, $\underline{d}^{(\alpha)'}$ satisfy the equation,

$$\underline{d}^{(\alpha)'} - \underline{d}^{(\alpha)} = \sum_{\beta=1}^{M_{\alpha}} \Delta^{(\alpha)}_{\beta}\,\underline{e}^{(\alpha)}_{\beta} \tag{33}$$

where,

$$\Delta^{(\alpha)}_{\beta} = \frac{M_{\alpha}+1}{M_{\alpha}}\ (d^{(\alpha)'}_{\beta} - d^{(\alpha)}_{\beta}) \tag{34}$$

Equation (33) means that one can construct a policy D' from another policy D by means of successive admissible variations $\delta_{\alpha\beta}(\Delta^{(\alpha)}_{\beta})$, $\alpha = 1,\ldots, N$; $\beta\varepsilon K_{\alpha}$. The variation in the expected average reward can be written as,

$$\begin{aligned}\phi(D') - \ (D) = {} & \frac{\partial\ \phi}{\partial d^{(1)}_{1}}\ (D)\ .\ \Delta^{(1)}_{1} + \frac{\partial\ \phi}{\partial d^{(1)}_{2}}\ (D+\delta_{11}(\Delta^{(1)}_{1}))\ .\ \Delta^{(1)}_{2} + \\ & +\ldots+ \frac{\partial\ \phi}{\partial d^{(1)}_{M_1}}\ (D+\delta_{11}(\Delta^{(1)}_{1}) +\ldots+ \delta_{1M_1}(\Delta^{(1)}_{M_1}))\ .\ \Delta^{(1)}_{M_1} + \\ & +\ldots+ \frac{\partial\ \phi}{\partial d^{(\alpha)}_{\beta}}\ (D + \sum_{\substack{\gamma<\alpha,\ \delta \\ \gamma=\alpha,\delta<\beta}} \delta_{\gamma\delta}(\Delta^{(\gamma)}_{\delta}))\ .\ \Delta^{(\gamma)}_{\delta} + \ldots. \\ & + \Delta^{(1)2}_{1}\ .\ \Psi_{11}(D) + \Delta^{(1)2}_{2}\ .\ \Psi_{12}(D+\delta_{11}(\Delta^{(1)}_{1})) + \ldots\end{aligned} \tag{35}$$

Since,

$$\frac{\partial \phi}{\partial d_\beta^{(\alpha)}}(D + \sum_{\substack{\gamma<\alpha,\ \delta \\ \gamma=\alpha,\delta<\beta}} \delta_{\gamma\delta}(\Delta_\delta^{(\gamma)})) - \frac{\partial \phi}{\partial d_\beta^{(\alpha)}}(D) \cong \sum_{\substack{\gamma,\delta \\ \gamma<\alpha \\ \gamma=\alpha,\delta<\beta}} \frac{\partial^2 \phi}{\partial d_\delta^{(\gamma)} \partial d_\beta^{(\alpha)}}(D) \cdot \Delta_\delta^{(\gamma)} \tag{36}$$

we get upon substituting into (35),

$$\phi(D') - \phi(D) = \sum_{\alpha,\beta} \frac{\partial \phi}{\partial d_\beta^{(\alpha)}}(D) \cdot \Delta_\beta^{(\alpha)} + \sum_{\alpha,\beta} \sum_{\substack{\gamma,\delta \\ \gamma<\alpha \\ \gamma=\alpha,\delta<\beta}} \frac{\partial^2 \phi}{\partial d_\delta^{(\gamma)} \partial d_\beta^{(\alpha)}}(D) \cdot \Delta_\delta^{(\gamma)} . \Delta_\beta^{(\alpha)} +$$

$$+ \sum_{\alpha,\beta} \Delta_\beta^{(\alpha)2} . \Psi_{\alpha\beta}(D) + o[(|\Delta_1^{(1)}| + |\Delta_2^{(1)}|+\ldots+ |\Delta_{M_N}^{(N)}|)^3] \tag{37}$$

THEOREM 1. Necessary conditions for local optimality of a policy D^* are

$$\text{i. } \frac{\partial \phi}{\partial d_\beta^{(\alpha)}}(D^*) = 0 \text{ all } \alpha : \underline{d}^{(\alpha)*} \in \text{int } S^M\alpha, \text{ all } \beta\varepsilon K_\alpha \tag{38}$$

$$\text{ii. } \frac{\partial \phi}{\partial d_\beta^{(\alpha)}}(D^*)\ \Delta_\beta^{(\alpha)} < 0 \text{ all } \alpha : \underline{d}^{(\alpha)*} \varepsilon\partial S^M\alpha, \text{ all } \beta\varepsilon K_\alpha$$

Proof. Suppose that condition i of the theorem is not satisfied for some $\alpha : \underline{d}^{(\alpha)*} \varepsilon$ int $S^M\alpha$, say for $\alpha = \bar{\alpha}$. Consider a variation $\delta_{\bar{\alpha}\beta}(\Delta_\beta^{(\bar{\alpha})})$ of the policy D such that $\Delta_\beta^{(\bar{\alpha})}$ is admissible. Since $\underline{d}^{(\bar{\alpha})*}$ is an interior point of the simplex $S^M\bar{\alpha}$ then $\Delta_\beta^{(\bar{\alpha})}$ can assume both positive and negative signs. Choose,

$$\text{sign } \Delta_\beta^{(\bar{\alpha})} = \text{sign } \frac{\partial \phi}{\partial d_\beta^{(\bar{\alpha})}}(D^*) \tag{39}$$

Hence for sufficiently small $\Delta_\beta^{(\bar{\alpha})}$ we get

$$\delta_{\bar{\alpha}\beta}\ \phi(D^*) > 0 \tag{40}$$

which contradicts the assumption that D^* is optimal. Hence condition i must hold. Now if condition ii does not hold for some $\alpha : \underline{d}^{(\alpha)} \varepsilon\partial S^M\alpha$, say for $\bar{\alpha}$, then again an admissible variation $\delta_{\bar{\alpha}\beta}(\Delta_\beta^{(\bar{\alpha})})$ will yield (40) which is in contradiction with the assumption that D^* is optimal.

THEOREM 2. Sufficient conditions for local optimality of a policy D^* are,

i. $\frac{\partial \phi}{\partial d_\beta^{(\alpha)}}(D^*) = 0$ all α : $\underline{d}^{(\alpha)*} \varepsilon$ int $S^M\alpha$, all $\beta \varepsilon K_\alpha$

ii. $\frac{\partial \phi}{\partial d_\beta^{(\alpha)}}(D^*) \cdot \Delta_\beta^{(\alpha)} < 0$ all α : $\underline{d}^{(\alpha)} \varepsilon \partial S^M\alpha$, all $\beta \varepsilon K_\alpha$ (41)

iii. $p_\gamma^* \xi_{\gamma\delta}^{\alpha *} = 0$ all α : $\underline{d}^{(\alpha)*} \varepsilon \partial S^M\alpha$, all γ : $\underline{d}^{(\gamma)*} \varepsilon \partial S^M\gamma$, all $\delta \varepsilon K_\gamma$

iv. $\sum_{\alpha:\underline{d}^{(\alpha)*} \varepsilon \text{int } S^{M_\alpha}} \sum_{k=1}^{M_\alpha} d_k^{(\alpha)*} \eta_{\alpha k} \xi_{\gamma\delta}^{\alpha *} \xi_{\gamma\delta}^{\gamma *} < 0$ or $p_\gamma^* = 0$ all γ:$\underline{d}^{(\gamma)*} \varepsilon$int $S^M\gamma$, all $\delta \varepsilon K_\gamma$

Proof. If the conditions of the theorem hold then it follows from the expansion formula (37) as well as the definitions (26), (28), and (32) that in the neighborhood of D^*,

$$\phi(D) - \phi(D^*) = \sum_{\substack{\alpha:\underline{d}^{(\alpha)*} \varepsilon \partial S^M\alpha \\ \text{all } \beta \varepsilon K_\alpha}} \frac{\partial \phi}{\partial d_\beta^{(\alpha)}}(D^*) \cdot \Delta_\beta^{(\alpha)} + \sum_{\alpha,\beta} \Delta_\beta^{(\alpha)2} \sum_{i:\underline{d}^{(i)*} \varepsilon \text{int } S^{M_i}} \sum_{k=1}^{M_i} d_k^{(i)*} \cdot \eta_{ik} \xi_{\alpha\beta}^{i*} \xi_{\alpha\beta}^{\alpha *} p_\alpha < 0 \quad (42)$$

for all admissible $\Delta_\beta^{(\alpha)}$. Hence D^* is an optimal policy.

4.3 AUTOMATON CONTROL MODEL.

i. Complete A priori Information. Let us consider a control decision model in the form of a stochastic automaton which experiments control policies while observing the system's state at the successive epochs 0, 1, 2, ... The automaton starts with an arbitrary policy D ; for example the randomised policy generating equally probable decisions for each observed state (i.e. $d_k^{(i)} = \frac{1}{M_i}$, all i, k). Let the present epoch be n and the automaton observe a state α of the system. According to the policy D(n) the automaton generates a control decision β with probability $d_\beta^{(\alpha)}(n)$; $\beta \varepsilon K_\alpha$. The system then makes a transition to a j-th state at the next epoch, n+1, and the automaton receives a reward $c_{\alpha\beta j}$. An "Adaptive Device", see Fig.1, changes the policy D of the stochastic automaton at n+1 in order to improve the expected average reward. The algorithm for changing the policy D, or what amounts to the same the structure of the automaton, is called a reinforcement scheme. That scheme is considered to be as follows. If at epoch n the observed state is α and the control decision is β then vary the policy at n+1 such that

$$\begin{aligned} \underline{d}^{(i)}(n+1) &= \underline{d}^{(i)}(n) + \underline{\delta}_{\alpha\beta}^{i}(\Delta_\beta^{(\alpha)}(n)), \; i=1,\dots,N \; ; \; k=1,\dots,M_i \\ \Delta_\beta^{(\alpha)}(n) &= \gamma_\beta^\alpha(n) \frac{\partial \phi}{\partial d_\beta^{(\alpha)}}(n) \quad , \quad \gamma_\beta^\alpha(n) > 0 \end{aligned} \quad (43)$$

subject to the bound (14).

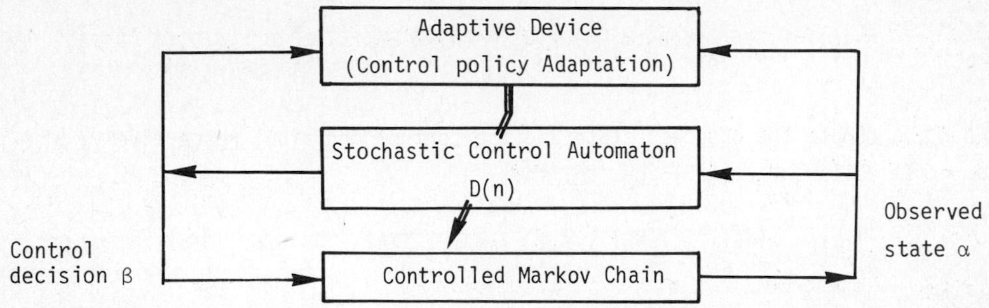

Fig.1 - Stochastic Control Automaton

ii. Lack of A priori Information. Let us consider the design of a control automaton which has to make control decisions without a priori knowledge of the transition probabilities $\pi(i,k,j)$. In that situation the automaton has to estimate the transition probabilities as time unfolds and simultaneously with decision making. In such case the variation $\Delta_\beta^{(\alpha)}(n)$ in (43) is given by

$$\Delta_\beta^{(\alpha)}(n) = \gamma_\beta^\alpha(n) \frac{\partial \tilde{\phi}}{\partial d_\beta^{(\alpha)}}(n) \quad , \quad \gamma_\beta^\alpha(n) > 0 \tag{44}$$

where $\partial \tilde{\phi}/\partial d_\beta^{(\alpha)}$ is the estimate of the gradient at epoch n.

Remark. If it happens that for some state i, at epoch n, the component $d_{k_0}^{(i)}$ of the policy $\underline{d}^{(i)}$ equals zero, and at the next recurrence of the state i the gradient corresponding to a decision $k \neq k_0$ is positive then the reinforcement scheme (43), or (44), will be applied on the simplex $S^{M_i-1} \subset S^{M_i}$ obtained by dropping off the k_0-th decision alternative. This is necessary to avoid premature convergence to a non-optimal policy.

4.4 CONVERGENCE.

Let the present epoch be n. Let $\mathcal{E}(n)$ denote the realization of the random variables : the decision probability vectors $\underline{d}^{(i)}(n')$, i=1,...,N, and the observed state x_n, for n' = 0,1,...,n. Consider the error criterion ; $I(n) = (\phi(n) - \phi^*)^2$ where, $\phi(n)$ is the expected average reward for policy $D(n) = D_n$ and ϕ^* is the optimal expected average reward. The expected value of I(n+1) conditioned on $\mathcal{E}(n)$ can be written as,

$$E(I(n+1)/\varepsilon(n)) = E((\phi(n) + \delta\phi(n) - \phi^*)^2/\mathcal{E}(n)) \tag{45}$$

$$= (\phi(n) - \phi^*)^2 + E((\delta\phi(n))^2/\mathcal{E}(n)) + 2E((\phi(n) - \phi^*)\delta\phi(n)/\mathcal{E}(n))$$

$$= I(n) + \sum_a (\delta_{x_n a}\phi(D_n,\Delta_a^{x_n}))^2 . d_a^{(x_n)}(n) + 2\sum_a (\phi(n)-\phi^*)\delta_{x_n a}\phi(D_n,\Delta_a^{x_n}) . d_a^{(x_n)}(n).$$

where x_n, a denote the observed state and the control decision, respectively, at epoch n. It follows from (25) and (26) that

$$\delta_{x_n a}\phi(D_n,\Delta_a^{(x_n)}) = \Delta_a^{(x_n)} . \frac{\partial \phi}{\partial d_a^{(x_n)}}(n) + (\Delta_a^{(x_n)})^2 \, f_{x_n a}(D_n,\Delta_a^{(x_n)}) \tag{46}$$

where,

$$f_{x_n a}(D_n, \Delta_a^{(x_n)}) = \sum_i \sum_k d_k^{(i)}(n) . \eta_{ik} \, . \, \xi^i_{x_n a}(n) . \frac{\xi^{x_n}_{x_n a}(n)}{1+\Delta_a^{(x_n)}\xi^{x_n}_{x_n a}(n)} . p_{x_n}(n) \tag{47}$$

and $\Delta_a^{(x_n)}$ is choosed according to the reinforcement scheme (44) as

$$\Delta_a^{(x_n)} = \gamma_a^{x_n}(n) \, . \, \frac{\partial \tilde{\phi}}{\partial d_a^{(x_n)}}(n) \tag{48}$$

Substituting into (46) we get

$$\delta_{x_n a}\phi(D_n,\Delta_a^{(x_n)}) = \gamma_a^{x_n}(n) \frac{\partial \tilde{\phi}}{\partial d_a^{(x_n)}}(n) \Bigg[\frac{\partial \phi}{\partial d_a^{(x_n)}}(n) + \tag{49}$$

$$+ \gamma_a^{x_n}(n) \frac{\partial \tilde{\phi}}{\partial d_a^{(x_n)}}(n) \, f_{x_n a}(D_n,\gamma_a^{x_n}(n) \frac{\partial \tilde{\phi}}{\partial d_a^{(x_n)}}(n) \Bigg]$$

Substituting by (49) into (45) we obtain,

$$E(I(n+1)/\mathcal{E}(n)) = I(n) + \sum_a (\gamma_a^{x_n})^2 \left(\frac{\partial \tilde{\phi}}{\partial d_a^{(x_n)}}(n)\right)^2 \Bigg[\frac{\partial \phi}{\partial d_a^{(x_n)}}(n) + \gamma_a^{x_n}(n) \frac{\partial \tilde{\phi}}{\partial d_a^{(x_n)}}(n) .$$

$$f_{x_n a}(D_n,\gamma_a^{x_n}(n) \frac{\partial \tilde{\phi}}{\partial d_a^{(x_n)}}(n)) \Bigg]^2 + \sum_a 2\gamma_a^{x_n}(n)(\phi(n) - \phi^*) \frac{\partial \tilde{\phi}}{\partial d_a^{(x_n)}}(n) \Big[\, . \, \Big] \tag{50}$$

Here $\Big[\, . \, \Big]$ means the same term between brackets as in the second terms of the same equation. We impose the condition that the estimated derivatives $\frac{\partial \tilde{\phi}}{\partial d_a^{(x_n)}}(n)$, following an appropriate estimation of the transition probabilities $\tilde{\pi}(i,k,j)$, satisfy the condition that

$$\frac{\partial \phi}{\partial d_a^{(x_n)}}(n) \left[\frac{\partial \phi}{\partial d_a^{(x_n)}}(n)+\gamma_a^{x_n}(n)\frac{\partial \tilde{\phi}}{\partial d_a^{(x_n)}}(n) f_{x_n a}(D_n,\gamma_a^{x_n}(n)\frac{\partial \tilde{\phi}}{\partial d_a^{(x_n)}}(n))\right] < \left(\frac{\partial \phi}{\partial d_a^{(x_n)}}(n)\right)^2 .$$

$$\cdot\ \rho_{x_n a}(n) + r_n - s_n \tag{51}$$

where,

$$\rho_{x_n a}(n) = 1+\gamma_a^{x_n}(n) f_{x_n a}(D_n,\gamma_a^{x_n}(n)\ \frac{\partial \phi}{\partial d_a^{(x_n)}}(n)) \tag{52}$$

r_n and s_n are non-negative $\mathcal{E}(n)$ - measurable random variables such that

$$\sum_n s_n < \infty \text{ a.s.} \quad ; \quad r_n, s_n \text{ are uniformly bounded.} \tag{53}$$

Let us examine the sign of $\rho_{x_n a}$, eqn. (52). Substituting from (47) we have,

$$\rho_{x_n a}(n)=1+\gamma_a^{x_n}(n) \sum_i \sum_k d_k^{(i)}(n)\eta_{ik}\ \xi^i_{x_n a}(n)\frac{\xi^{x_n}_{x_n a}(n)}{1+\gamma_a^{x_n}(n)\frac{\partial \phi}{\partial d_a^{(x_n)}}(n)}\ p_{x_n}(n) \tag{54}$$

Using the boundedness conditions (11), (22) as well as the definition of the first-order derivatives (26) we get,

$$\frac{\partial \phi}{\partial d_a^{(x_n)}}(n) < (\max_a \eta_{x_n a} - \min_a \eta_{x_n a})+c(x_n)\max_{x_n, a} \eta_{x_n a} < c_1+c_o c_1 = c_2$$

$$\frac{\partial \phi}{\partial d_a^{(x_n)}}(n) > (\min_a \eta_{x_n a} - \max_a \eta_{x_n a})-c(x_n)\max_{x_n, a} \eta_{x_n a} > -c_1-c_o c_1 =-c_2 \tag{55}$$

Hence $\rho_{x_n a}(n)$ is a uniformly bounded sequence for which

$$\rho_{x_n a}(n) \geq 1 - \bar{\gamma}\ \frac{c_1 c_o^2}{1-\bar{\gamma}\ c_2} \tag{56}$$

where $\bar{\gamma}$ is an upper-bound to be imposed on the sequence $\gamma_a^{x_n}(n)$. It follows then that $\rho_{x_n a}(n)$ is a non-negative sequence if

$$\bar{\gamma} \leq \frac{1}{c_2 + c_1 c_o^2} \tag{57}$$

Having guaranteed the non-negativeness of the variables $\rho_{x_n a}$, eqn.(54), we can rewrite eqn.(50) in the form of the inequality

$$E(I(n+1)/\mathcal{E}(n)) < I(n) + \bar{c} \sum_a (\gamma_a^{x_n}(n))^2 - \sum_a 2\, \gamma_a^{x_n}(n)(\phi^* - \phi(n)) \Big[\Big(\frac{\partial \phi}{\partial d_a^{(x_n)}}(n)\Big)^2 \rho_{x_n a}(n) + r_n - s_n \Big] \tag{58}$$

where $\bar{c}$ is a positive constant representing the uniform upper-bound,

$$\bar{c} \geq \max_{\tilde{\pi}, d_a^{(x_n)}, \gamma_a^{x_n}} \Big(\frac{\partial \tilde{\phi}}{\partial d_a^{(x_n)}}\Big)^2 \left[\frac{\partial \phi}{\partial d_a^{(x_n)}} + \gamma_a^{x_n} \frac{\partial \tilde{\phi}}{\partial d_a^{(x_n)}} f_{x_n a}\Big(D_n, \gamma_a^{x_n} \frac{\partial \tilde{\phi}}{\partial d_a^{(x_n)}}\Big) \right]^2 \tag{59}$$

Hence $I(n)$ is a non-negative almost supermartingale and we can apply the convergence theorem of Robbins and Siegmund[3]. This yields the following result. $\lim_n I(n)$ exists and is finite and

$$\sum_{n=1}^{\infty} \sum_a 2\, \gamma_a^{x_n}(n)(\phi^* - \phi(n)) \left[\Big(\frac{\partial \phi}{\partial d_a^{(x_n)}}(n)\Big)^2 \rho_{x_n a}(n) + r_n \right] < \infty \tag{60}$$

on

$$\sum_a (\gamma_a^{x_n})^2 < \infty \tag{61}$$

If we further impose the conditions

$$\sum_{n=1}^{\infty} \gamma_a^{x_n}(n) = \infty \quad , \quad \sum_{n=1}^{\infty} \gamma_a^{x_n}(n)\, r_n < \infty \tag{62}$$

then (60) combined with the fact that $\Big(\frac{\partial \phi}{\partial d_a^{(x_n)}}\Big)^2 \rho_{x_n a}$ is a uniformly bounded positive sequence ; cf.(55), and (56), imply that

$$\lim_n \phi(n) = \phi \quad \text{w.p.1} \tag{63}$$

We summarize the above results in the following theorem.

<u>THEOREM 3</u>. The reinforcement scheme (44) subject to the conditions (51), (53) as well as

$$0 \leq \gamma_a^{x_n} \leq \bar{\gamma} = \frac{1}{c_2 + c_1 c_o^2}, \quad \sum_{n=1}^{\infty} \gamma_a^{x_n}(n) = \infty, \quad \sum_{n=1}^{\infty} (\gamma_a^{x_n}(n))^2 < \infty, \quad \sum_{n=1}^{\infty} \gamma_a^{x_n}(n) r_n < \infty$$

yields an expected average reward with probability one.

4.5 ACCELERATION.

The following idea, originally due to Kesten[4], may be employed to accelerate the convergence of the stochastic algorithm (44). When the policy D is far from the optimal there will be few changes of sign of successive values of the gradient $\partial\phi/\partial d_\beta^{(\alpha)}$; $\alpha = 1,\ldots,N$, $\beta = i,\ldots,M$. Near the optimal, we would expect to cause oscillation from one side of $d_\beta^{(\alpha)*}$ to the other. This suggests using the number of sign changes of successive values of $\partial\phi/\partial d_\beta^{(\alpha)}$ to indicate whether the policy estimate $d_\beta^{(\alpha)}$ is near or far from $d_\beta^{(\alpha)*}$. To accelerate convergence, the quantity $\gamma_\beta^\alpha(n)$, see (44), is not decreasing if $\partial\phi/\partial d_\beta^{(\alpha)}$ has the same sign as the respective preceding value (i.e. for the same α, β). To formalize this, we introduce the set of N vectors $\underline{Z}^{(1)},\ldots,\underline{Z}^{(N)}$. If at epoch n the event : state α, decision β, is taken place then the β-th component of the vector $\underline{Z}^{(\alpha)}$ will be defined as,

$$Z_\beta^{(\alpha)}(n) = \text{sign}\,\frac{\partial\,\phi}{\partial d_\beta^{(\alpha)}}(n), \quad \text{sign}\,(0) = 0 \tag{64}$$

We also introduce the set of N count vectors $\underline{L}^{(1)},\ldots,\underline{L}^{(N)}$ which are initialized as

$$L_j^{(i)}(0) = 0, \quad j = i,\ldots, M_i\;;\; i = 1,\ldots,N \tag{65}$$

If at epoch n the event : state α, decision β, is taken place then the β-th component of the vector $\underline{L}^{(\alpha)}$ will be up-dated thus,

$$L_\beta^{(\alpha)}(n) = L_\beta^{(\alpha)}(n-1) + 1 \tag{66}$$

The step-length vectors $\underline{\gamma}^i = (\gamma_1^i,\ldots,\gamma_{M_i}^i)$, $i = 1,\ldots,N$ are first initialized as,

$$\gamma_j^i(0) = \gamma_o = \text{const.} > 0, \quad j = 1,\ldots, M_i,\; i = 1,\ldots, N \tag{67}$$

If at epoch n the event : state α, decision β is taken place then the step-length element $\gamma_\beta^\alpha(n)$ is defined as,

$$\gamma_\beta^\alpha(n) = \begin{cases} \gamma_\beta^\alpha(n-1)\,, & \text{if } (L_\beta^{(\alpha)}(n) < 2)\cup((L_\beta^{(\alpha)}(n) \geq 2)\cap(Z_\beta^{(\alpha)}(n).Z_\beta^{(\alpha)}(n-1) > 0) \\ \dfrac{\gamma_o}{L_\beta^{(\alpha)}(n)}\,, & \text{if } (L_\beta^{(\alpha)}(n) \geq 2)\cap(Z_\beta^{(\alpha)}(n).Z_\beta^{(\alpha)}(n-1) \leq 0) \end{cases} \tag{68}$$

The sequence $\gamma_j^i(n)$ must satisfy the condition that the respective policy increments $\Delta_j^{(i)}(n)$ satisfy the constraint (14) in order to verify that $\underline{d}^{(i)}(n+1)$ belongs to the simplex S^{M_i}. If $\Delta_j^{(i)}(n)$ is such that $\underline{d}^{(i)}(n+1)$ does not belongs to the simplex then $\gamma_j^i(n)$ is divided by two. The process of division is repeated, if necessary, until $\underline{d}^{(i)}(n+1)$ is found to be in the simplex S^{M_i}.

4.6 NUMERICAL EXAMPLE.

We consider the example of the "Taxicab operation" given by Howard[5]. The problem consists of a taxicab driver whose territory encompasses three towns A, B and C. If he is in town A, he has three alternatives :

1. He can cruise in the hope of picking up a passenger by being hailed.
2. He can drive to the nearest cab stand and wait in line.
3. He can pull over and wait for a radio call.

If he is in town C, he has the same three alternatives, but if he is in town B, the last alternative is not present because there is no radio cab service in that town. For a given town and given alternative, there is a probability that the next trip will go to each of the towns A, B and C and a corresponding reward in monetary units associated with each such trip. This reward represents the income from the trip after all necessary expenses have been deducted. For example, in the case of alternatives 1 and 2, the cost of cruising and of driving to the nearest stand must be included in calculating the rewards. The probabilities of transition and the rewards depend upon the alternative because different customer population will be encountered under each alternative.

If we identify being in towns A, B and C with states 1, 2 and 3, respectively, then we have Table 1.

Table 1. Data for Taxicab Problem.

State	Alternative	Probability			Reward			Expected Immediate Reward
i	k	$\pi(i,k,j)$			r_{ikj}			r_{ik}
1	1	1/2	1/4	1/4	10	4	8	8
	2	1/16	3/4	3/16	8	2	4	2,75
	3	1/4	1/8	5/8	4	6	4	4,25
2	1	1/2	0	1/2	14	0	18	16
	2	1/16	7/8	1/16	8	16	8	15
3	1	1/4	1/4	1/2	10	2	8	7
	2	1/8	3/4	1/8	6	4	2	4
	3	3/4	1/16	3/16	4	0	8	4,5

Simulation of the stochastic control algotithm (43) is carried out on a digital computer. Starting from a completely random policy (i.e. all decisions are made with equal probabilities) the policy converged w.p. 1 to the optimal one, see Fig.2. The long-run expected average reward always increased at every epoch until it reached the steady state optimal value, see Table 2.

Table 2.

epoch n	expected average reward ϕ
0	9.179964
1	9.680186
2	12.465918
3	12.650611
4	12.660767
5	12.817069
10	13.177875
20	13.241918
30	13.318928
40	13.342069
50	13.344534

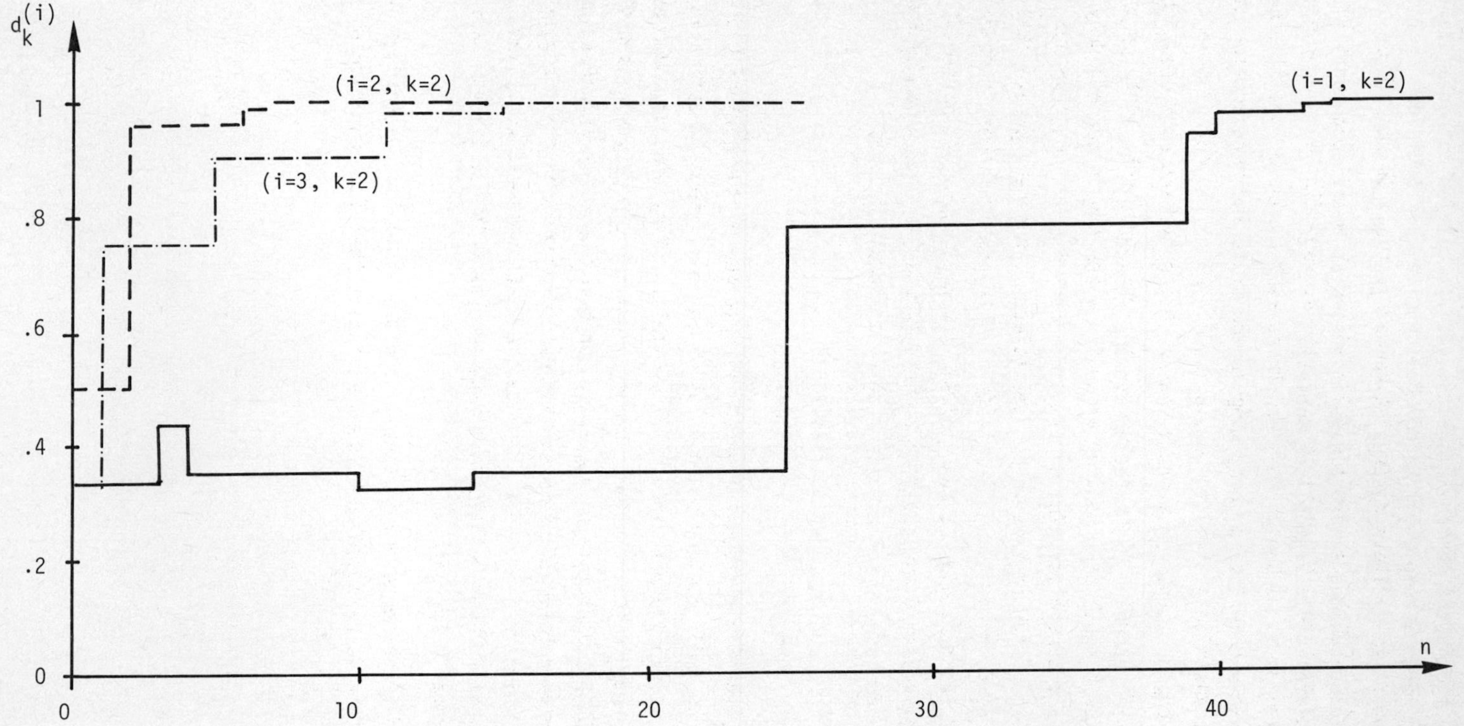

Fig.2 - Control policy versus time

4.7 CONCLUSIONS.

The control of a finite-state discrete-time Markov chain is considered. At each epoch the state of the chain is observed and a control decision has to be taken. Possible control decisions are finite for all states. Depending on the observed state and the taken decision the chain makes transition to one of the alternative states with certain probabilities. Depending on the state and the decision as well as the effected transition a specific reward is obtained. The control objective is to maximize the long-run expected average reward. Necessary and sufficient conditions of optimality are established using a variational approach. That approach allows to formulate a stochastic control algorithm which can be performed by a stochastic automaton controller. The automaton chooses its actions (control decisions) for each state with respective probabilities. It has been proven that the automaton's decisions converge with probability 1 to the optimal ones. A numerical example is worked out on a digital computer and the results are in agreement with the theory. The algorithm is believed to be versatile as it uses a little computation time and memory and enjoys a non-negative super-martingale property.

COMMENTS.

4.2 The idea of using a variational approach is inspired by the work of Lyubchik and Poznyak[6]. They provided a sketchy formulation of the conditions of optimality for the optimal control problem with inequality constaints. They did not, however, indicate any concrete meanings of the "derivatives". Neither did they evaluate the nature of their conditions (sufficient, necessary, or both). Further work remains to be done for the problem with constraints, which may be formulated as a stochastic programming problem[7]. Theorems 1, and 2 presented here are believed to be new. It is interesting to examine their relationship with the Howard's conditions[5] based on the dynamic programming approach.

4.3 The presented convergence proof is new. Condition (51) for the case of lack of a priori information is believed to be less stringent than the condition of minimum contrast estimate, cf. Mandl[8].

REFERENCES

1. D.L. Isaacson, and R.W. Madsen, Markov Chains. New York : John Wiley and Sons, 1976.

2. E. Durand, Solutions Numériques des Equations Algébriques, Tome II. Paris : Masson 1961.

3. H. Robbins, and D. Siegmund, "A convergence Theorem for Non-negative Almost Super-Martingales and Some Applications", in Optimization Methods in Statistics, ed. by J.S. Rustagi. New York : Academic, 1971.

4. H. Kesten, "Accelerated Stochastic Approximation", Ann. Math. Statistics, 29, 1, pp. 41 - 59, 1958.

5. R.A. Howard, Dynamic Programming and Markov Processes, New York : John Wiley and Sons, 1962.

6. L.M. Lyubchik, and A.G. Poznyak, "Learning Automata in Stochastic Plant Control Problems", Automation and Remote Control, N°6, pp. 777 - 789, 1974.

7. A.S. Poznyak, "Learning Automata in Stochastic Programming Problems", Automation and Remote Control, N°10, pp. 1608 - 1619, 1973.

8. P. Mandl, "Estimation and Control in Markov Chains", Adv. Appl. Prob. 6, pp. 40 - 60, 1971.

EPILOGUE

In Chapter I we have discussed existing definitions of learning and delineated new aspects of cybernetic modeling of what can be a learning or self-organizing system. We highlighted the dynamics of real system-environment interactions during a learning process. Further work is needed to carry over those ideas to the realm of real self-organizing systems. That would need first to introduce the role of energy and matter into the information perspective and moreover find formulae for their metabolism and transformation into one another. That would surpass the limits we have deliberately settled for our quest in this monograph. Chapter II has grown out from a survey work of the basic techniques to pattern recognition systems. It has brought together a host of published results in a unified, systematic, and critic way. Chapter III presents a complete theory of learning automata use to solve collective problems, which are formulated as a game between automata. It has been shown that under certain conditions the individualistic goal-seeking behavior of the automata converges to a desired meta-objective set out by the environment. Further research can prove to be fruitful to study situations where the environment is non-stationary in contradistinction with the stationarity hypothesis considered in the present work. Chapter IV presents a new algorithm for the control of finite Markov chains with unknown transition probabilities. A theory is given ensuring the convergence w.P.1. Further research is to be developped to study cases where constraints exist on the control and the state. Also to investigate the possibility of using a team of learning automata to deal with chains with a high number of states, a situation where the "curse of dimensionality" prevails.